Edward Jackson

Skiascopy and its Practical Application to the Study of Refraction

Edward Jackson

Skiascopy and its Practical Application to the Study of Refraction

ISBN/EAN: 9783337407582

Printed in Europe, USA, Canada, Australia, Japan

Cover: Foto ©berggeist007 / pixelio.de

More available books at **www.hansebooks.com**

SKIASCOPY

AND ITS

Practical Application to the Study of Refraction

BY

EDWARD JACKSON, A. M., M. D.,

PROFESSOR OF DISEASES OF THE EYE IN THE PHILADELPHIA POLYCLINIC AND COLLEGE FOR
GRADUATES IN MEDICINE; SURGEON TO WILLS EYE HOSPITAL; EX-CHAIRMAN OF THE
SECTION ON OPHTHALMOLOGY OF THE AMERICAN MEDICAL ASSOCIATION;
MEMBER OF THE AMERICAN OPHTHALMOLOGICAL SOCIETY;
ETC., ETC.

SECOND EDITION

WITH TWENTY-SEVEN ILLUSTRATIONS

COPYRIGHT, 1896
BY EDWARD JACKSON

CONTENTS

PREFACE .   5

CHAPTER I.
- History   7
- Name   10
- Difficulties   11
- How to Study the Test   12

CHAPTER II.—GENERAL OPTICAL PRINCIPLES.
- The reversal of movement   18
- The Point of Reversal   20
- Real Movement of light on retina, Plane Mirror . .   21
- Real Movement of light on retina, Concave Mirror . .   23
- Apparent Movement of light in pupil   25
- Rapidity of Movement of light on the retina . . .   27
- Magnification of the retina   29
- Form of the light area   30
- Brightness of light in the pupil   32
- Finding the point of reversal   33

CHAPTER III.—CONDITIONS OF ACCURACY.
- The Source of light   35
- Focusing of light on the retina   37
- Position of observer for greatest accuracy . . .   39
- Irregularities of the Media and Surfaces . . .   40
- Distance of Surgeon from Patient . . , . .   42
- Final test   44

CHAPTER IV.—REGULAR ASTIGMATISM.
- Two points of reversal   45
- The Band-like appearance   46
- Changes in the Light area at different distances . .   51
- Direction of movement of the band   53

CHAPTER V.—ABERRATION AND IRREGULAR ASTIGMATISM.
- Appearance of Irregular Astigmatism.   55
- Symmetrical Aberration   57
- The Visual Zone   58
- Appearances of Positive Aberration   59

Appearances of Negative Aberration . . 62
Appearance of Conical Cornea 64
Scissors-like Movement 66

CHAPTER VI.—PRACTICAL APPLICATION WITH PLANE MIRROR.

Position and Arrangement of Light 69
Hyperopia 70
Myopia 72
Emmetropia . . . , ... 74
Regular Astigmatism 75
Aberration and Irregular Astigmatism 82
Measurement of Accommodation 84

CHAPTER VII.—PRACTICAL APPLICATION WITH CONCAVE MIRROR.

The Source of Light. 87
Hyperopia 88
Myopia . . . . , ... 89
Emmetropia 91
Regular Astigmatism 91
Aberration and Irregular Astigmatism 96
Measurement of Accommodation 97

CHAPTER VIII.—GENERAL CONSIDERATIONS.

Apparatus 98
Mydriatics 104
Relative advantages of plane and concave mirrors . 105

INDEX 107

PREFACE

THE sale of the first edition of this book within one year has shown a general interest in its subject among working ophthalmologists, and gives reason to hope that it will, to some extent, accomplish its object.

It was written to bring about the more general adoption of Skiascopy as an essential part of the examination for ametropia. It is not supposed that any ophthalmologist is quite ignorant of the test; but many do not know its full practical value, or how best to apply it.

The demonstrations and descriptions here given assume a general knowledge of the eye and of physiological optics. And the writer, having observed that students of this subject do not generally think in the terms of algebraic formulas, but more readily grasp the graphic or geometric presentation of a fact, has resorted to the latter method so far as practicable.

The claims of this subject to careful consideration are:

FIRST.—Skiascopy, is an objective test, independent of the patient's intelligence or visual acuteness, and more largely than any other, independent of the patient's cooperation.

SECOND.—It is by far the most accurate objective test. The limits of its accuracy depend on details of its execution, and the skill and patience of the observer; but, it does not require any rare natural qualifications, to carry it, for many eyes, to the extreme limits of accuracy for subjective tests.

THIRD.—It requires but little more time than the use of the refraction ophthalmoscope or the ophthalmometer, which are able to give very inferior information. It saves time in making a complete diagnosis.

FOURTH.—It requires no costly, complex or cumbersome apparatus.

FIFTH.—It lays before the surgeon the refraction in each particular part of the pupil as it is revealed by no other test, opening up the principal avenue for farther advance in the scientific study of the refraction of the eye.

Of the use of the data obtained by means of skiascopy, it is not the purpose of the present monograph to speak. These data include the refraction of the visual zone, corresponding to the refraction of the eye obtained by other methods; an accurate knowledge as to the location and limits of that zone; and the refraction outside of it; the latter having in some cases important bearings on the practical adjustment and use of lenses.

The changes made in this edition while numerous are most of them slight, being chiefly alterations of phraseology intended to render more clear the expression of the thought. One new illustration has been added.

Philadelphia, June, 1896.

CHAPTER I.

HISTORY, NAME, DIFFICULTIES, AND METHOD OF STUDYING THE TEST.

History.—From the earliest use of the ophthalmoscope it has been recognized that, by the direct method, the seeing of an erect image of the fundus at some little distance from the eye indicated hyperopia, and the seeing of an inverted image indicated myopia; and that the distance of the inverted image from the eye indicated the degree of myopia. So long ago as 1862, Bowman (*The Royal London Ophthalmic Hospital Reports,* Vol. II, p. 157), called attention to the rotation of the mirror as a means of bringing out appearances characteristic of irregular astigmatism and conical cornea, and Donders, in his work on *Accommodation and Refraction of the Eye,* published in 1864, includes (p. 490) the following note:

"My friend Bowman recently informs me that 'he has been sometimes led to the discovery of regular astigmatism of the cornea, and the direction of the chief meridians, by using the mirror of the ophthalmoscope much in the same way as for slight degrees of conical cornea. The observation is more easy if the optic disc is in the line of sight and the pupil large. The mirror is to be held at two feet distance, and its inclination rapidly varied, so as to throw the light on the eye at small angles to the perpendicular, and from opposite sides in succession, in successive meridians. The area of the pupil then exhibits a somewhat linear shadow in some meridians rather than in others.'"

The use of the ophthalmoscope above referred to, for the detection of irregular astigmatism, became widely popular. It was generally adopted as the most satisfactory test for this kind of defect. But the observation that the same method was capable of revealing regular astigmatism and the direction of its principal meridians, does not seem to have attracted general attention.

In 1872, Couper, in his paper before the Fourth International Ophthalmological Congress (see *Trans*. page 109), alluded to Bowman's observations, and said: "The greater dispersion in one meridian than in the opposite, gives rise to the linear shadows. Only the fact of astigmatism is thus established." He then went on to describe a method of using the ophthalmoscope as an optometer in astigmatism, which is rather a modification of the ordinary use of the ophthalmoscope than a variety of skiascopy, since it depends on the recognition or non-recognition of the retinal vessels in different meridians, when the ophthalmoscope mirror is held at a considerable distance from the eye, and takes no account of the movement of a light area on the retina.

In 1873, Cuignet, of Lille, published (*Rec. d'Ophtalmol.*, 1873, pp. 14 and 316) an account of the test, as he had used it, as one capable of revealing not only the presence of hyperopia or myopia as well as astigmatism, but also as giving a practical method of measuring the amount of these errors of refraction. He seems not to have appreciated fully the optical principles involved in the test, and his account of it attracted no attention. However, in 1878, his pupil, Mengin, introduced the practice of the method at Galezowski's clinic in Paris. There it was taken up, and Parent demonstrated its true optical basis and urged its advantages in a series of articles published in the *Recueil d'Ophtalmologie* in 1880–81, pp. 65 and 229.

Lytton Forbes, in the *Royal London Ophthalmic Hos-*

pital Reports for 1880, (p. 62), published a paper on the test, giving a minute account of the various forms assumed by the light and shadow in the pupil, but without full explanation of their optical significance. In 1881, A. Stanford Morton included a full description of the test in his little work on the *Refraction of the Eye*. In 1882, Charnley gave the fullest demonstration of its optical basis in the *Royal London Ophthalmic Hospital Reports*, X, 3, p. 344. And Juler called attention to it in the *Ophthalmic Review*, Vol. I, p. 327.

The method described and advocated by Parent and those who followed him, had been that with the concave mirror. Cuignet had used the plane minor, and, in 1882, Chibret pointed out (*Annales d'Oculistique*, Vol. XXXVIII, p. 238) the advantages of the plane mirror in determining the presence and degree of myopia in the examination of large numbers of recruits. In 1883, Story (*Ophthalmic Review*, Vol. II, page 228) advocated the use of the plane mirror, but in the same manner as the concave, except that the observer should place himself at a distance of four metres from the patient, a distance which renders the test of little value for a considerable proportion of cases.

In 1885, was published (*American Journal of the Medical Sciences*, April, p. 404) the writer's account of the test with the plane mirror, as applicable to all varieties of ametropia, the determination being made by measuring the variable distance of the surgeon from the patient. Since that time the literature of the subject has grown rapidly and the value of the test has been widely recognized; but even yet it is far from being universally adopted and depended upon as it deserves to be. It must be noted that a considerable proportion of the accounts of the test bear evidence that their authors' acquaintance with it has been theoretical rather than practical, and the mass of them contribute nothing to the common fund of professional knowledge. The writer's contributions as to the retinal illumination (*Ophthalmic Review*, Feb., 1890) and the relative positions of the source of light and the observer (*Archives of Ophthalmology*, July, 1893), with the numerous suggestions of others as to the special pieces of apparatus to facilitate the test, which will be mentioned in Chapter VIII, complete the evolution of this method of diagnosis as now practiced.

Name of the test.—Neither Bowman nor those who after him early employed the test for the detection of irregular astigmatism and conical cornea, proposed for it any special name.

Cuignet, who brought it forward as a distinct method for the diagnosis of the refraction of the eye, seems to have thought at first that the play of light and shade in the pupil depended entirely on the curvature of the cornea, and described it under the name *keratoscopie*. Considering the real causes of the movement of light and shade in the pupil and the purposes for which it is employed, this name seems especially inappropriate.

Parent, realizing this inappropriateness, proposed *retinoscopie*, in allusion to the fact that it was the movement of light and shade on the pigment layer of the retina that commonly gave rise to the phenomena studied. Yet this name was obviously open to criticism, in that the condition of the retina itself was not at all the matter under consideration ; and that the same play of light and shade could be watched on the head of the optic nerve, or, where they were exposed, upon the choroid or sclera.

Chibret, to bring out the point that it was the movement of a shadow that was the subject of investigation, proposed the name of *fantoscopie retinienne,* and, Mr. Priestley Smith, probably anglicizing this term and dropping the allusion to the retina, called it the *shadow-test*. This name though a compound word, where a simple one should do, became extremely popular, and its appropriateness led Chibret to call to his aid, the linguistic skill of M. Egger, who rendered it in the term *skiascopia,* which in its French form *skiascopie,* or its English form, *skiascopy,* has been most widely accepted as the proper term to designate the test.

Umbrascopy, proposed by Hartridge, is indefensible on linguistic grounds, and the same is true of *pupilloscopie* proposed by Landolt, and for which he aftewards offered the equivalent, *koroscopie. Dioptroscopie* was advocated by Galezowski (*Atlas d'Ophthalmoscopie*) and is appropriate, though equally applicable to other methods of measuring refraction.

Retinophotoscopie and *retinoskiascopie* have been more recently suggested by Parent, but there seems to be no sufficient reason for retaining in the name any allusion to the retina. *Fundus-reflex-test* suggested by Oliver is also unnecessarily long for a name.

The suggestion has sometimes been made to apply one of these names to one form, and another name to another form of the test. But such a use of them is not warranted by the intention of their proposers or by custom. Nor is there any sufficient reason for the employment of separate names to differentiate various forms of the test; in all its different forms, the test is essentially the same, the difference being merely as to the apparatus and mechanical detail.

Difficulties of the Test.—That skiascopy, though a valuable method of examination, is one difficult to completely master, becomes more and more evident as one continues to work with it. The theoretical basis is perfectly simple, the fundamental phenomena readily observed; and, with a few days practice, the merest tyro may be able by it to estimate the refraction in favorable eyes with an accuracy not to be attained by any other objective method. But long after the stage of such acquirement has been passed, the surgeon will again and again encounter cases that still prove difficult and puzzling. Nothing but a thorough understanding of the optical principles involved, and patient study of the eyes which prove most puzzling, under carefully arranged favorable conditions, will enable him to master the test.

The importance of the careful arrangement of the relative positions of the light and of the observer, and the adaptations of the mirror have not heretofore been sufficiently insisted upon. What these adaptations and arrangements are will appear under their proper headings in chapters III and IV. It is here only necessary to emphasize their importance. For instance: All descriptions of the shadow-test allude to the characteristic band-like appearance of the light in astigmatism. Now, as a matter of fact, even in the highest degrees of astigmatism, such an appearance cannot be perceived, except with certain lenses, or at certain distances in front of the eye; and it is a distinctive and exact indication only when the light, the mirror, and the patient's and the observer's eyes are brought into a certain relation.

It would be as rational to attempt to measure refraction with an ophthalmoscope devoid of any lens series, or to test the acuteness of vision in a darkened room, as to expect definite and satisfactory results from skiascopy, applied without careful attention to details that have often not been referred to in descriptions of the test.

The fact that this test shows, as does no other, the actual refraction of the eye for each particular portion of the pupil, increases enormously the wealth of phenomena it offers for study, adding to its scientific and practical value, but also making it more difficult by rendering it necessary to discriminate between the particular portions of the movements of light and shade which are of practical importance, and others which are not.

How to Study the Test.—The study of skiascopy is something quite different from its practical application. To start from a few bare rules as to the placing of glasses, and the movements of the mirror, and the light in the pupil; and attempt to learn the test by using it will never give a mastery of it. It is better to make a careful study of it before attempting to employ it as a method of ascertaining the refraction.

Such a study is chiefly a use of the test, but from a standpoint entirely different from that of its application in practice. To study the test, one should as far as possible, start with known conditions of refraction, with lenses of known strength, with the eye at a known distance, and should observe the character of the movements of light and shadow in the pupil, which belong to these known conditions. In studying it, one should work from known refraction to the pupillary appearances caused by it; while in using the test for the measurement of ametropia, he has to deduce from observed pupillary appearances the state of refraction causing them.

The student may, from time to time, test his progress towards proficiency by attempts to measure refraction by skiascopy, but at first familiarity with the appearances indicative of known conditions of refraction is chiefly to be sought.

The appearances upon which the attention is fixed in skiascopy are those of the red reflex in the pupil. The first step is to learn just what the reflex in the pupil is and some of the variations which it may exhibit. Let the beginner, with his eye at the sight-hole of the skiascopic mirror throw into the observed eye, from a distance of 20 or 30 inches, the light from a lamp flame, as in the ordinary ophthalmoscopic examination. Looking into the observed eye with the light properly directed, he will see the brilliant point of light, the reflection from the surface of the cornea of the lamp flame he is using; and he may also see reflections of his own face or of other objects from the surface of the cornea. These are to be disregarded. The real object of study, the phenomena upon which attention is to be fixed, is the general red glow perceived within the pupil, the fundus reflex.

If the mirror be rotated about an axis lying in the plane of the mirror, the area of light thrown by it upon the face will move in the direction towards which the mirror is turned. As the test becomes familiar, the direction of this movement will be known without any conscious effort to discover it. With the concave mirror at a greater or lesser distance than its focus or with the plane mirror at all distances, except at the point of reversal which it is the object of the test to determine, the rotation of the mirror also causes a movement of the red reflex in the pupil. As the reflex passes off across the pupil, it is followed by an area of shadow, and, as it returns across the pupil, the shadow passes out before it. The movement of the light area really goes on when no shadow is visible in the pupil, but only when light and shade are both seen can the movement be recognized. We know the movement of light in the pupil by the movement of the boundary between light and shade.

Having learned what it is that he has to watch in the pupil, the student should make himself familiar with the various appearances of the fundus reflex and its movements, by viewing it from different distances, with different lenses before the eye, with different mirrors, and later in a number of different eyes; and all this without, at first, concerning himself especially as to the state of refraction that causes the particular appearance that he sees. That is, he should first learn to some extent what are the variations in the pupillary reflex. A few of them are illustrated in the following pages. A second step will be the attempting to appreciate their significance.

Without a good understanding too of the simple optical

principles underlying the test, it must remain a blind routine and rule of thumb work, and can never be of the highest utility. To aid in such an understanding of them, one may, in connection with the study of the succeeding chapters, take a strong (15 D. to 20 D.) convex lens and a piece of card-board with a dot on it. The lens can represent the dioptric media of the eye, the card-board the retina, and the dot the light area upon the retina. The card-board should be held back of the lens, a little farther than its focal distance, and the dot looked at through the lens from various distances. Nearer the lens an erect image of the dot (blurred of course), and, farther away, an inverted image will be seen, and between the two the phenomena of reversal. The movement of light on the retina may be imitated by a slight movement of the card in different directions.

The apparent enlargement of the dot, as the point of reversal is approached, and the diminution of its apparent size as the point of reversal is departed from, its diffusion and indistinctness near the point of reversal, and its concentration and greater definiteness away from the point of reversal, are to be observed. Such a combination of dot and lens will also beautifully exhibit the phenomena of aberration [See Chap. V] with its central and peripheral areas of differing movement, the one an erect and the other an inverted image. The difficulty of keeping the dot in view when the point of reversal is approached, will illustrate how small a portion of the retina is visible from the point of reversal when the test is applied to the eye. By holding in combination with the spherical lens a cylindrical lens of 5 D., the distortions of the light area produced by astigmatism, and the band-like appearances it causes at certain distances, should also be studied.

This is not all to be done at a single lesson, but the lens and card should be kept at hand where they can be used to imitate and elucidate the different conditions as they arise in studying the pupillary reflex.

The study of the appearances in the eye may thus be carried on: Take an eye, the refraction of which is known, and from a distance that will give an erect movement, throw the light into the eye, and, by the rotation of the mirror, produce and study the erect movement. Then with a lens which it is known will give an inverted movement, the inverted movement is to be similarly studied. Finally the lens, or position of the observer is to be so varied as to bring the point of reversal to the observer's eye, and the appearance of the pupil from this point is also to be studied. In these studies, and, indeed, throughout the whole course, the student will find it easier first to master and understand the appearances with the plane mirror.

If it is possible to get an eye free from astigmatism and aberration of any notable degree, the earlier studies of the pupillary appearances will be much simplified. After the appearances in such an eye have become familiar, the phenomena of astigmatism may be studied by placing before the same eye, a cylindrical lens of known strength. The point of reversal with such a lens will give the observer the appearances presented by the pupil at that distance; and at other distances from the eye the other appearances presented in astigmatism can be obtained.

For example, suppose the eye at the student's disposal to be hyperopic 1 D. Let him first place before it the convex 2 D. lens. This will bring the point of reversal one metre from the eye. With the plane mirror, let him first study the erect movement at one-half metre; then study the inverted movement at a distance of two metres; then observe the eye from the point of reversal at one metre, and then vary his distance so as to study it from intermediate points.

When he takes up the study of astigmatism, he should

place before such an eye, a convex cylindrical lens of 2 D. in addition to the spherical. Then from the distance of one-third of a metre he will be able to observe the band of light at right angles to the axis of the lens, from a distance of one metre the band of light running in the direction of the axis of the lens, and from other distances the other appearances indicative of astigmatism.

Familiarity with the many appearances due to aberration and irregular astigmatism will only be obtained by study of eyes presenting those defects. But, as the great majority of eyes present them in notable degree, material for such a study is not difficult to obtain. Careful observation of the corresponding appearances, with the lens and card-board already referred to, will enable the beginner promptly to recognize the appearances of aberration. And, when once he has found an eye that presents them, let him carefully study them with the plane mirror, with different lenses, and from various distances.

A considerable part of the study of skiascopy and especially of the appearances of positive aberration, can be carried on with the aid of an artificial, schematic, or model eye. That of Frost is one of the best, although any, even the rudest, will answer. In the studies on the human eye, it is better to study one eye long and repeatedly, or at most to confine the earlier observations to a few eyes than to attempt to employ a large number. Each additional eye will introduce variations in the appearances presented, which will at first be only puzzling and retard, rather than assist, the mastery of the test.

CHAPTER II.

GENERAL OPTICAL PRINCIPLES. MOVEMENT, FORM AND BRIGHTNESS OF LIGHT AREA.

Skiascopy is a method of measuring myopia, either the myopia originally present in the eye or that produced by a lens of known strength for the purpose of measurement. In myopia, we have the retina situated back of the principal focus of the dioptric media, so that rays of a certain divergence, that is coming from a point a certain finite distance in front of the eye, are brought to a focus upon the retina. Conversely, the rays coming from a point of the retina and passing out through the crystalline lens and cornea, are brought to a focus at the same distance in front of the eye. The point for which the eye is focused, and the point on the retina, on which the focused rays are received, have to the refractive surfaces of the eye the relation of conjugate foci.

The Reversal of Movement.—The amount of myopia is known when we know the distance of the point in front of the eye, which has this relation of a focus conjugate to the retina. Skiascopy furnishes a method of determining the position of this point. Closer to the eye, than this point for which it is focused, the observer may see an erect image of the fundus. Farther from the eye than this point, he can perceive an inverted image. Skiascopy is a means of determining when the image seen is erect and when it is inverted, or when it passes from the erect to the inverted.

When this occurs may be understood from a study of figure 1, Let M represent a myopic eye, A and B being two points of the retina from which rays emerge to reach the observer's eye ; and C and D the points at which these rays coming from the retina are focused, the rays coming from A being focused at C and those from B at D.

The apparent position of a point is determined by the direction of a ray coming from that point, and passing through the nodal point of the observer's eye. Suppose the observer's eye is placed at N, closer than the point for which the observed eye is focused. The apparent position of the point A is determined by a ray which passes through the upper part of the pupil and is turned down. It appears

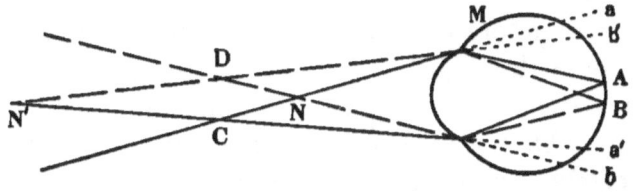

FIG. I.

in the direction of a. The apparent position of the point B will be located by the ray coming through the lower part of the pupil and turned up. It will be seen in the direction of b. Thus, from this position N, the point, A which is really above appears above, and the point B, which is really below appears below. The observer sees an erect image.

When, however, the observer places his eye at N', at a greater distance than that for which the eye is focused, the ray which reaches his nodal point from A, will be one that comes through the lower part of the pupil and is turned up; so that A will appear to be located in the direction of a' in the lower part of the pupil. From this position he will judge the location of B by the ray which comes through the upper part of the pupil and is turned down, so that B will appear to be located in the direction of b' in the upper part of the pupil. That is, the point A, which is really above, will appear to be below, and the point B, which is really below will appear to be above. The image observed is inverted.

The Point of Reversal.—It is evident that this change in the relation of the rays, that brings about the change in the apparent position of A and B, occurs at the distance of the points C and D, at which, the rays coming from the retina are focused. Here it is that these rays intersect and take their new relation which gives the reversal of the apparent position of the points of the retina from which they come.

It is, therefore, convenient in connection with skiascopy to designate this point as *the point of reversal.* The name indicates the significance of this point with reference to this test. Of course, it is really the same point as the far point of the myopic eye—the point for which the eye is focused—the conjugate focus of the retina—these latter names indicating the relations of the same point in other matters.

It is only when the rays leave the eye convergent, only when the eye is myopic, that they ever come to a focus in front of it. If the eye be emmetropic or hyperopic, the rays emerging parallel or divergent remain so at all distances. Hence, in emmetropia and hyperopia, there can be no point of reversal. From whatever distance the eye is viewed, the image perceived is erect.

In myopia, the distance of the point of reversal from the eye depends on the degree of convergence of the rays as they leave the cornea—depends on the amount of myopia. The distance of the point of reversal from the eye, being the distance from the eye to its far point, is the focal distance of the lens required to correct the myopia. So that to ascertain the amount of myopia, we have only to determine the point of reversal, and measure its distance from the eye.

Skiascopy determines the position of the point of reversal by observation of the directions of the movement of light and shade in the pupil. Other kinds of ophthalmoscopic examinations attempt the recognition of the details of the fundus image. But, as the point of reversal is approached, the details of the fundus image become indistinct and fade away entirely, so that the location of the point of reversal cannot be accurately determined by such an examination. On the other hand, when this point has been so closely approached that the fundus details are quite indistinguishable, it still remains easy to recognize the direction of the movement of light and shade in the pupil ; and, from it, to deduce the erect or reversed character of the image. Skiascopy, therefore, determines the point of reversal, and measures the degree of myopia with much greater exactness than Couper's or the fundus image test.

In skiascopy, we watch the *apparent* movement of light and shade in the pupil, due to the *real* movement of an area of light upon the retina. This area of light is secured by reflecting into the eye the light from a lamp with a skiascopic mirror. This is done in a darkened room, in order that the retina outside of this light area may be dark, furnishing a decided contrast to the area to be watched. The movement of the light area upon the darkened retina is secured by varying the inclination of the mirror, rotating it about some axis lying in the plane of the mirror and passing through the sight hole. The direction of the movements thus produced by a certain change in the inclination of the mirror depends on whether it is plane or concave.

Real Movement of the Light on the Retina. The Source of Light.—The lamp flame, or similar source of light used for the test, may be called the *orignal source* of light, in contra-distinction to the reflection of it from the mirror, which being more immediately related to the movement of the light on the retina, we shall call the *immediate source* of light.

22 GENERAL OPTICAL PRINCIPLES.

The Plane Mirror.—With the plane mirror the immediate source of light is behind the mirror as far as the original source of light is in front of it. The rays reflected from the mirror enter the eye under observation as though they had started from this immediate source. As the mirror is rotated, the apparent position of the immediate source of light changes; for this immediate source is situated upon a line drawn through the original source perpendicular to the surface of the mirror, and necessarily changes with that perpendicular as the inclination of the mirror changes.

With the change of position of the immediate source of light, the rays coming from it and falling upon the eye, are made to fall upon a new part of retina, and thus the inclination of the mirror causes a change in the part of the retina that is lit up by the light reflected into the eye.

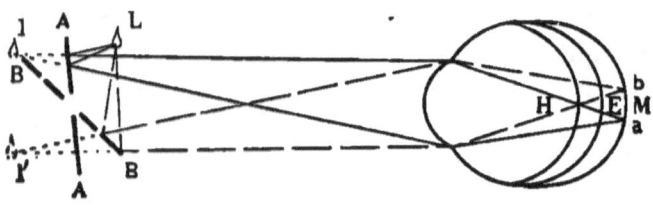

FIG. 2.

What these changes are can be better understood by a study of figure 2. L represents the position of the lamp flame, the original source of light. When the mirror is held in the position AA, the immediate source of light is situated at l, and light entering the eye from that direction falls upon the retina toward a. When, however, the position of the mirror is changed to BB, the immediate source of light is changed to l', from which, light falls upon the retina toward b. As the mirror is rotated from AA to BB, the position of the immediate source of light moves from l to l', and, as a consequence, the area of light upon the retina moves from *a* to *b*. The light on the retina then, moves in the direction that the mirror is made to face. It is said to move *with the mirror.*

Only a portion of the light reflected by the mirror enters the eye, the remainder falls upon the face and makes an area of light on the face. One may readily demonstrate by trial that this area of light cast by the mirror on the face also moves *with* the mirror under all circumstances.

The rays of light coming from *l* and *l'* intersect at the nodal point of the eye; and passing directly on do not again change their relative positions. Whatever the distance of the retina from this nodal point, the movement of the light upon it will be in the same direction, so that whether the retina be at H. as in hyperopia, at E. as in emmetropia, or at M. as in myopia, the *real movement* of light upon it from a certain movement of the mirror is always in the same direction.

Therefore, *with the* PLANE MIRROR, *the* REAL MOVEMENT *of the area of light on the retina is* WITH *the mirror—with the area of light on the face—in all states of refraction.* This is true for all distances of the light from the mirror, or of the light and mirror from the tested eye.

The Concave Mirror.—With the concave mirror as used in skiascopy, the immediate source of light is a real focus of the mirror, conjugate to the position of the light, situated between the mirror and the eye to be tested. The position of this immediate source varies with the position of the mirror, moving in the direction that the mirror is made to face and causing an opposite movement in the area of light that falls from it upon the retina.

In figure 3 *L* again represents the original source of light. When the mirror is in the position *AA*, the light falling upon it from *L* is focused at *l.*, and the little inverted image of the lamp flame there formed is the immediate source of light. From it the rays diverge, some to fall upon the face, and those entering the eye to fall upon the retina toward a. When the mirror is turned to occupy the position BB, the light falling upon it is focused at l', which

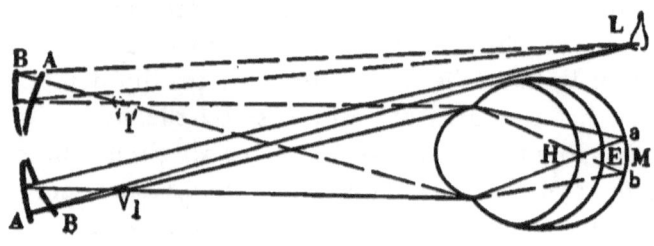

Fig. 3.

becomes the new position of the immediate source of light, and from which the rays entering the eye fall upon the retina toward b. As the mirror is rotated from AA to BB, the immediate source of light moves from l to l' and the light upon the retina from a to b. This will be the direction of its movement in all states of refraction whether the retina be situated at H. as in hyperopia, at E. as in emmetropia, or at M. as in myopia. The portion of the light which falls upon the face, however, and forms the facial area, as can be readily demonstrated by trial, moves in the direction that the mirror is made to face.

We have then: *With the* CONCAVE MIRROR, *the* REAL MOVEMENT *of the area of light on the retina is* AGAINST *the mirror, and against the light on the face, in all states of refraction. And when, in the following pages, "erect" or "direct" movement is spoken of with the concave mirror, such movement "against" the mirror is meant.*

The above is the movement that occurs with the concave mirror used as in skiascopy so far from the original source of light and from the eye to be tested, that the conjugate focus of the original source of light falls in front of the eye. If, however, the original source of light be brought so close to the mirror that the rays from it are not rendered convergent, but continue to diverge after reflection, the immediate source of light will be a magnified image of the lamp flame, situated behind the mirror as in the case of the plane mirror; and the movement of the retinal light area will be precisely the same as with the plane mirror. Again, if the rays reflected by the mirror are rendered convergent, but the eye to be tested is brought so near that they cannot come to a focus in front of its nodal point, the light will pass in as though from an immediate source back of the mirror, and the movement of the area of light on the retina will again be like that with the plane mirror. If the light reflected upon the eye be convergent so as to be focused just at its nodal point, no movement of light on the retina, such as we have been considering, will occur; but whatever direction the mirror is turned, so long as the light enters the eye, the retinal light area will remain stationary.

It is to be borne clearly in mind that the movement so far spoken of is the *real* movement of the light area upon the retina, as it would appear from within the eye itself, or when viewed from behind the retina, with the sclera and choroid cut away.

The Apparent Movement of the Light in the Pupil.—What we observe in skiascopy, however, is the apparent movement of the light in the pupil as viewed from the position of the observer some distance in front of the eye. When an *erect image* of the retina is viewed, this *apparent movement* of the light will be in the *same direction* as the *real movement*. When an *inverted image* is viewed, the *apparent movement* will be in the *direction opposite* to that of the *real movement*.

The observer can always watch the movement of the light area on the face, and know that with the plane mirror the light area on the retina always has a *real* movement in the same direction, and with the concave mirror it always has a *real* movement in the opposite direction; and he has only to compare the *apparent* movement of the light which he watches in the pupil with the known direction of the real movement on the retina, to determine whether he sees an erect or an inverted image. When the apparent and real movements are in the same direction, he knows (page 18) he is looking at the eye from a distance shorter than that for which it is focused. When the apparent and real movements are in the opposite directions, he knows that he is looking at the eye from a distance greater than that for which it is focused.

The direction of the apparent movement of the light then, will be with the light on the face in hyperopia and in emmetropia at all distances, and in myopia when the eye is viewed from a point nearer than its point of reversal. The apparent movement in the pupil will be the opposite of the real movement only in cases of myopia when the eye is viewed from somewhere beyond its point of reversal.

With the PLANE MIRROR, *the apparent movement is* WITH *the light on the face in hyperopia, emmetropia, and myopia with the point of reversal behind the observer, and* AGAINST *the light on the face in myopia viewed from beyond the point of reversal.*

With the CONCAVE MIRROR *the apparent movement is* AGAINST *the light on the face in hyperopia, emmetropia, and myopia with the point of reversal behind the observer; and is* WITH *the light on the face only in myopia viewed from beyond the point of reversal.* This statement made to conform to the practice customary in the use of the concave mirror [where the observer keeps a constant distance of 1 metre from the eye, corresponding to 1 D. of myopia], would be: *the light moves* AGAINST *the light on the face and against the mirror in hyperopia, emmetropia, and myopia of less than 1 D., and only moves* WITH *the light on the face in myopia of more than 1 D.*

When we speak of "inverted" movement with the concave mirror, this movement with the mirror and with the light on the face is meant.

These statements have been made with reference to the apparent movement of the light before the state of refraction has been modified by any glass placed before the eye for that purpose. But they hold equally as to hyperopia, emmetropia, or myopia remaining uncorrected or produced by a lens placed before the eye. For instance:—In myopia the movement remains *against* the light on the face with the plane mirror, or *with* the light on the face with the concave mirror, so long as the concave lens employed is not strong enough to bring the point of reversal to the distance of the observer's eye. In hyperopia or emmetropia, where the movement is watched through a convex lens, the movement remains *with* the light on the face for the plane mirror, and *against* the light on the face for the concave mirror, until a convex lens is used, that is strong enough to over-correct the hyperopia and cause enough myopia to bring the point of reversal nearer to the eye than the position of the observer.

Rapidity of Movement of the Light on the Retina.
—The rapidity with which the light and shadow appear to move across the pupil depends first, on the rapidity of the real movement of the light area upon the retina; and, second, upon the magnification of the retina. The rapidity of the real movement on the retina depends:

On the rate of movement of the mirror in the observer's hand.

On the distance of the mirror from the observed eye.

On the distance of the original source of light from the mirror.

And upon the distance of the retina from the nodal point in the observed eye.

The rate of movement of the mirror and the distance of the light from the mirror determine the rapidity of the movement of the immediate source of light; this being greater as the mirror is moved more quickly, or as the original source of light is more distant from the mirror. The excursion which the immediate source of light can make is limited by the width of the mirror, and the extent of movement of the light area on the retina produced by the movement of the immediate source of light entirely across the mirror depends on the relative distance of the mirror and the retina from the nodal point of the eye. The wider the mirror, or, the nearer it is to the nodal point of the eye, or the farther the retina is from that nodal point, the greater the extent of movement produced in the retinal area of light by a given movement of the mirror. On account of the relative distances of the retina from the nodal point, the extent of the movement of the light on the retina is, other things being equal, least in the highest hyperopia and greatest in the highest myopia.

The *rapidity* of the *real movement* of the light on the retina then, *is increased:*

By moving the mirror faster.

By carrying the original source of light farther from the mirror.

By bringing the mirror closer to the eye.

By elongation of the antero-posterior axis of the eyeball.

The *real movement* of the light upon the retina is *made slower:*

By moving the mirror more slowly.

By bringing the original source of light closer to the mirror.

By carrying the mirror farther from the eye.

By shortening of the antero-posterior axis of the eyeball.

In using the test, the distance of the light from the mirror is, for most purposes, practically constant, and the ordinary variations in the antero-posterior axis of the eyeball are so slight as to have no appreciable influence. So that the rapidity of the *real* movement of light on the retina depends principally on the rapidity of the movement of the mirror and the distance of the mirror from the eye.

Magnification of the Retina.—In practice the rapidity of the *apparent* movement of the light in the pupil depends far more on the extent to which the retina, and the real movement of light upon it are magnified, than upon the actual rate of that real movement. The retina, as viewed through the pupil from different distances, is seen under different degrees of magnification. When the observer's eye is placed at the point of reversal, the rays from a single point of the retina, passing through all parts of the pupil, converge to the observer's nodal point, so that the one point of the retina appears to occupy the whole of the pupil, and the retina is seen indefinitely magnified. As the observer's eye departs from the point of reversal, it receives the rays from an increasing area of the retina, more and more of the retinal image occupies the same space of the pupil and the retina is seen less magnified.

This is illustrated in figure 4, which represents an eye with its point of reversal at *A*. If the observer's eye be placed at *A* it receives rays only from the point *a*, and this point appears to occupy the whole pupil. If, however, the observer's eye be placed at *B*, from which rays would be

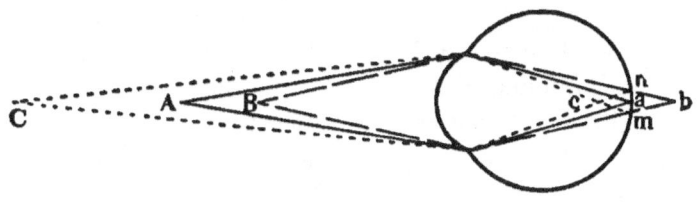

FIG. 4.

focused at *b* behind the retina, and, at which, rays from *b* would be focused, the observer will be able to see in the space of the pupil all of the retina, *m n* included, between the broken lines passing from *B* to *b*—all of the retina, which would receive a circle of diffusion if the rays were coming from the point *B*. Or, again, if the observer's eye be placed at *C*, from which rays will be focused at *c* in front of the retina, and, at which, rays coming from *c* would be focused, he will be able to perceive the portion of the retina, *m n* included, between the dotted lines, passing through *c* and continued on to the retina—the area upon which would be formed a circle of diffusion by rays coming from the point *C*.

It follows then that *the closer the observer's eye to the point of reversal, the more is the real movement of light upon the retina magnified, and, therefore, the swifter does it appear. The farther the observer's eye is removed from the point of reversal, the less is that real movement of light on the retina magnified; and the slower is the apparent movement as watched in the pupil.*

And, as this source of variation overcomes all other sources of variation in the rate of the apparent movement of the light [except the rate of rotation of the mirror, which is, to a considerable extent, under the control of the observer], *the rapidity of the apparent movement of light and shade in the pupil increases as the point of reversal is approached, and diminishes as that point is departed from*, and constitutes a measure of the degree of ametropia remaining uncorrected.

Form of the Light Area.—The real form of the light area on the retina, except under certain conditions in astigmatic eyes, will be circular. If the light be perfectly focused on the retina it is circular, because that is the form of the source of light employed (see Chapter III). If the light be not perfectly focused on the retina, the circular pupil gives its form to the resulting area of diffusion.

The influence of regular astigmatism on the apparent form of the light area as seen in the pupil will be discussed in Chapter IV; and the influences of irregular astigmatism and aberration in Chapter V. These influences, especially the latter, are really dominant, and of the greatest practical importance in nearly all eyes. But in eyes free from such defects the form varies with the departure of the observer's eye from the point of reversal. If the magnification of the

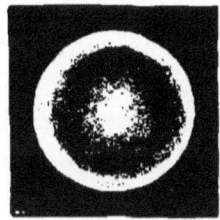
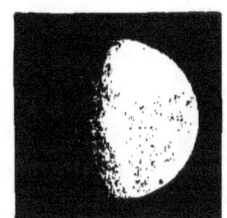

FIG. 5.   FIG. 6.

retina is so slight that all of it occupied by the light area is visible in the pupil at one time, that area appears circular, as represented in figure 5. But when the point of reversal is approached so that the magnification of the retina prevents all of the retinal light area from being seen at one time, only a portion of its outline is visible as an arc of the greatly enlarged circle, as shown in figure 6; and the nearer to the point of reversal that the observer comes, the nearer does the boundary between light and shade approach to a straight line. It must be borne in mind, however, that this is still part of the boundary of a circle, and hence that different parts will run in all the different directions, in contradistinction to the band-like appearance of astigmatism, the direction of which always conforms to one or the other of the principal meridians.

From the point of reversal, however, but a single point of the retinal light area could be visible to the observer at a time, so that the form of that area could not from this position influence the form of light and shade apparent in the patient's pupil. From this single luminous point of the patient's retina the light is not focused on the observer's retina, but falls there in an area of diffusion which would take its form from the pupil of the observer, or rather from the sight-hole of the mirror, which is the smaller. This being circular would give a circle of diffusion, the portion of which was referred to the patient's pupil, giving its shape to the apparent light area. In practice, however, we do not produce just these conditions. For reasons referred to under the next heading we do not commonly watch the movement of light and shade from exactly the point of reversal. Still, as this point of reversal is approached, this form of the diffusion area on the observer's retina often exerts some influence on the apparent form of light and shade in the patient's pupil. This may be demonstrated by trial with a square or triangular sight-hole. For a full discussion of this matter the reader is referred to a paper by Dr. Carl Weiland in the *Medical News*, October 12th, 1895; and one by the author *Annals of Ophthalmology and Otology*, April, 1896.

Brightness of the Light in the Pupil.—This depends on the illumination of the retinal light area and the extent to which that area is magnified.

The illumination of the light area on the retina depends on the brightness of the original source of light and the accuracy with which the light coming from it is focused on the retina. The brighter the source of light and the more accurately it is focused, the brighter the illumination of the retina. The dimmer the light and the larger the circle of diffusion over which it is dispersed, the more feeble the retinal illumination.

As the immediate source of light is usually near the mirror (in front for the concave, behind for the plane) when the mirror and the observer's eye approach the point of reversal, or the point of reversal is brought to them by a change of lenses, the light is more nearly focused on the retina, and the *actual illumination* of the light area in the patient's eye brighter.

But, as the point of reversal is approached, the *appar-*

ent brightness of the light area in the pupil is diminished by the increasing magnification of the retina, which causes the light from a smaller part of the retinal area to occupy the whole space of the pupil. Again, when the observer is near the point of reversal, the part of the retina that he can see is the part to which light can be reflected only from the immediate vicinity of the sight-hole and the sight-hole itself. On this account the illumination of this part of the light area is feeble as compared with the illumination of other parts of the light area ; and if the refraction throughout the pupil were uniform there would be a central portion in complete darkness. This is, however, prevented by the astigmatism and aberration, present in all eyes, which prevent the perfect focussing of the light even when the immediate source of light is exactly at the point of reversal.

On this account the brightest apparent illumination of the pupil is never obtained at the point of reversal, but usually at one or two dioptres from it, the exact position being dependent on the arrangement of the source of light.

Finding the Point of Reversal.—The point of reversal is to be recognized only when the observer's eye is in its immediate neighborhood. This may be effected either by varying the distance of the observer's eye from the observed eye until it comes to the position of the point of reversal, or by varying the position of the point of reversal by changes in the lenses placed before the observed eye until the point of reversal comes to the chosen position of the observer's eye. For reasons to be stated in Chapter VI, the former method is the better when using the plane mirror, and the latter is to be resorted to when the concave mirror is employed. In any case, the trial movement across the pupil shows by the direction of the movement whether a point of reversal exists between the observer and the observed eye, and the rapidity of movement shows approximately [when the observer has learned to appreciate its significance] the extent of the interval between the position of the observer and the point of reversal. If the movement be slow, the interval is large, perhaps several dioptres. If it be rapid, the interval is less.

Upon these data of the direction and rapidity of the movement, the surgeon bases the next step of the test, the selection and placing of lenses before the eye. This being done, the test is repeated, the movement seen through the lens noted, both as to its direction and rapidity, and the distance of the observer from the patient, or the strength of the lens before the observed eye, varied in accordance therewith. This process is continued until the observer's eye reaches the point of reversal, or the point of reversal is brought by the lens to the observer's eye. But *the test should not be regarded as complete until the movement has been repeatedly viewed both from within and from beyond the point of reversal*, as well as from that point. Only by this precaution of observing from a slightly greater and a slightly less distance, or with a slightly stronger or slightly weaker lens than that which brings the point of reversal to the surgeon's eye, can the certainty of a correct result be assured.

CHAPTER III.

CONDITIONS OF ACCURACY.

Since in skiascopy one has to observe the movement of an area of light across the shaded retina, the size, brightness and sharpness of the contrast between the margin of this light area, and the shadow immediately adjoining it are very important factors in determining the definiteness and accuracy of the test. For reasons discussed in the preceding chapter, the contrast between light and shadow as seen in the pupil necessarily diminishes as the point of reversal is approached. It is, therefore, important to have the contrast between light and shadow upon the retina as sharp as possible.

Darkening the Room.—To secure this contrast, the retina outside of the proper light area should be in absolute darkness. This requires a complete darkening of the room in which skiascopy is practiced, including the shading of the source of light, except in the direction in which it is used. The difference in the ease of the test applied in a completely darkened room, as contrasted with its use in a partially darkened room, can only be appreciated by one accustomed to applying it under the former condition.

The Source of Light.—To secure the brilliant illumination of the light area, in contrast with the complete shadow around it, the source of light must be as bright as possible. On account of the difficulty about the sight hole to be referred to later, the arc electric light cannot be employed, except to illuminate a piece of ground glass as suggested by Derby. The incandescent electric light is not available on account of its form, so that recourse must be had to one of the various illuminating flames. Of these the acetylene flame is most brilliant. Next come paraffin, the heavy mineral oils, and gas flames reinforced with the richer hydro-carbons, or used on the Welsbach mantle, and after this the ordinary illuminating gas. But a good flame of the latter furnishes a satisfactory illumination.

It is more important, whatever flame is used, that the brightest part of it should be employed. With all flames there is at the margin a comparatively gradual shading from light to darkness, which interferes with the sharpness of the boundary of the light area on the retina. To secure that sharp boundary as well as to prevent the diffuse illumination of the room, and to limit the size of the source of light, the flame should be entirely covered by an opaque shade with an aperture of the proper size placed opposite the most brilliant part of the flame. This gives, under proper conditions of focusing, a perfectly sharp margin to the light area on the retina.

The *size* of this opening in the opaque screen determining the size of the original source of light is governed by various conflicting requirements. The smaller the source of light, the more characteristic its shape in regular astigmatism, and the easier to distinguish the different movements in different parts of the pupil in irregular astigmatism and aberration. But enough light must be furnished by it to give a distinct area of light upon the face, as well as to give sufficient illumination within the pupil; and the source of light must be considerably larger than the sight hole in the mirror. As the mirror is rotated, the immediate source of light appears to move across it, and if this source were not larger than the sight hole, it would, at times, entirely disappear within that opening. At such times, no light would fall upon the retina and the illumination would disappear entirely from the pupil, causing delay and uncertainty in the test. If, however, the immediate source of light is larger than the sight hole, no such disappearance of the light occurs.

The size of the opening through which light is obtained, is then a compromise between the requirements of light and the size of the sight hole on the one hand, and a need to have the retinal light area as small as possible on the other. The diameter of the source of light for accurate work with the plane mirror may generally be reduced to little more than twice the diameter of the sight hole. In practice the writer prefers an aperture five millimetres in diameter, but the beginner may find one of double that diameter more satisfactory.

The Sight Hole.—This must be large enough to allow the observer to readily watch through it the movement of light on the face, as well as in the pupil. This will be facilitated by the thorough darkening of the room, and the complete freedom of the sight hole from reflections. With this limitation, the sight hole should be as small as possible, to reduce the area from which little or no light is reflected into the eye, and to reduce the circles of diffusion formed on the observer's retina when he watches the movement of light and shade from the immediate vicinity of the point of reversal. A sight hole two millimetres in diameter has rendered the writer the best service.

Focusing of the Light on the Retina.—When the rays coming from the immediate source of light are accurately focused upon the retina, the area of retinal illumination will be the smallest and brightest, and will have the most definite edge. This accurate focusing is secured only when the *immediate* source of light is situated at the focus conjugate to the retina—the point of reversal. In searching for the point of reversal, it is, therefore, advantageous to keep the immediate source of light as close to the observer's eye and the mirror as possible.

With the *plane mirror* the immediate source of light is a reflection of the original source as far behind the mirror as the immediate source is in front of it. The closer the original source of light can be brought to the mirror, the closer will its reflection be to the observer's eye; and to the point of reversal, at the critical moment when the observer's eye reaches that point. The *original* source of light then should be kept as close to the mirror as possible. On this account it should be moveable, to follow the movements of the observer's eye and the mirror, when the distance of these from the eye under observation is varied.

When the observer withdraws to the distance of two metres or more from the patient, it may not be practicable to keep the light very close to the mirror, but at such a distance, the separation of the source of light from the mirror becomes of small importance. For, if the original and immediate sources of light were at the mirror, the rays from the latter would have a divergence of one-half dioptre when they reached the eye; and, if the original source of light were one metre in front of the mirror, so that the immediate source would be one metre behind the mirror, that is three metres from the eye, the rays from it would reach the eye one-third dioptre divergent, and the difference between the one-half and the one-third dioptre of divergence is so trifling as to be in this connection of no practical importance.

On the other hand, when the surgeon approaches close to the patient's face, the slight distance that must necessarily remain between the original source of light and the mirror becomes a source of imperfect focusing of the light on the retina, and therefore of inexactness in the determination of the point of reversal. Suppose the mirror to be at five inches from the eye and the original source of light three inches from it, this will make the immediate source of light eight inches from the eye, and the rays from it will reach the pupil 5 D. divergent when the surgeon is seeking the point of reversal corresponding to 8 D. of myopia. This difference of 3 D. interferes greatly with the delicacy of the test.

With the *concave mirror*, the *immediate* source of light being a real image of the *original* source in front of the mirror, at its focus conjugate to the position of the original source, cannot be brought closer to the mirror than its principal focal distance. It is brought closest by carrying the original source of light as far away from the mirror as possible. The original source of light then, for the concave mirror, should be behind the patient as far as possible.

The important exception to these rules for placing the light, necessary for the most accurate determination of the principal meridians in regular astigmation is discussed in connection with that subject in Chapter IV.

Position of Observer for Greatest Accuracy.—With the *plane mirror* the *immediate* source of light is necessarily behind the mirror. It will, therefore, be exactly at the point of reversal when the mirror and the observer's eye are slightly within the point of reversal. Hence the conditions of accuracy are better complied with for the observation that is made from within the point of reversal, where the light still moves in the pupil with the light on the face, than for the observation that is made from beyond the point of reversal, where movement is inverted. The point of reversal is then, with the plane mirror, most closely approximated from the side toward the observed eye; and in practice the greatest accuracy is attained by considering that *the point of reversal is located at the greatest distance from the eye at which erect movement can be seen in the visual zone of the pupil.*

With the *concave mirror* the *immediate* source of light being necessarily in front of the mirror, can be brought accurately to the point of reversal only when that point of reversal is the focal distance of the mirror in front of the observer's eye. The point of reversal then is approached more accurately with the lens which still leaves it in front of the observer's eye, than with the lens which removes it back of the observer's eye. Hence, *with the concave mirror, the strongest concave lens, or the weakest convex, which allows the movement of the light in the pupil with the light on the face, is the lens which brings the point of reversal most accurately to the distance chosen.*

In regular astigmatism, as will be shown in chapter IV, the position of the light must be specially arranged, when it is desired to develop the band-like appearance characteristic of that condition. For the measurement of refraction in either of the principal meridians, the adjustment should be precisely the same as for simple hyperopia or myopia. But the band-like appearance cannot certainly be recognized unless the necessary conditions there explained, as to the position of the observer and source of light, are carefully observed. When the proper precautions are taken one can get a characteristic band of light with even one-half dioptre of astigmatism, and by that band can fix the direction of the principal meridians with great accuracy.

In the higher degrees of regular astigmatism there is considerable difficulty in measuring the refraction of the principal meridians with accuracy. It is, therefore best, before regarding the skiascopic test as completed, to place before the eye such a cylindrical lens as appears to be required to correct the astigmatism, repeat the test, and so ascertain whether the astigmatism has been accurately corrected. Fuller references to this matter will be found in the Chapters VI and VII.

Irregularities of the Media and Surfaces.—These interfere with skiascopy not only by changes in the apparent movement of the light as watched in the pupil, but also by preventing the perfect focusing of the light which falls upon the retina, and in this way, they limit to some extent, the accuracy of the test, since they are present in some degree in nearly all eyes.

In the case of positive aberration (see Chapter V), the interference with focusing is of the same kind as the defect in the refraction of a strong convex spherical lens. If one takes such a lens and intercepts with a piece of card-board the narrowing pencil of rays that have passed through the lens, he will find that the strong refraction at the margin of the lens causes a ring of condensation at the periphery of the circle of diffusion. This ring is exhibited from close behind the lens back to its principal focus, beyond which, we have the condensation at the centre of the light area and a gradual fading away of light around it. Hence, the circle of diffusion in front of the principal focus presents a brilliantly illuminated edge in sharp contrast with the shadow around it, while at the principal focus and behind it, the light area has not a sharply defined edge, but fades gradually into the shadow around it.

Therefore, in making the test, the influence of positive aberration upon the distribution of light in the light area may be utilized by having the light focused not exactly on the retina, but slightly back of it. This may be brought about by having the immediate source of light closer to the observed eye than are the observer's eye or the point of reversal; conditions that are secured in the use of the concave mirror. Hence, for positive aberration of a certain distribution in the pupil, a more definitely bounded light area might be obtained by the use of the concave mirror than could be had with the plane mirror. But the disadvantage of an indistinct margin of the retinal light area due to aberration is partly or entirely balanced by the obliteration from that area by this same aberration of the unilluminated or poorly illuminated centre due to the sight-hole.

With negative aberration, where the refraction is weaker near the periphery of the pupil, the condensation ring of light is less pronounced and is found back of the principal focus for the central visual area. For this form of aberration the plane mirror, enabling the observer by pushing the source of light from the mirror to get the light focused in front of the retina, has some advantage over the concave mirror.

The interference with the focusing of the light on the retina due to irregular astigmatism cannot be overcome in any way, and it impairs the value of the test and makes it more difficult to apply in eyes presenting marked defects of this kind.

Distance of the Surgeon from the Patient.—It will always be impossible to determine the point of reversal with perfect exactness. The best that can be done is to make out that it lies, within narrow limits of possible error, at about a certain distance. It may be a little nearer, it may be a little farther off.

If the distance be a short one, if the lens used be such that the point of reversal is brought close to the observed eye, the probable inaccuracy of distance will cause an appreciable error in estimating the refraction, measured in dioptres. For instance: at eight inches from the eye, two inches additional, making ten inches, will correspond to a whole dioptre of refraction, and two inches less, making the distance six inches, will correspond to a difference of a dioptre and a half. On the other hand, at eighty inches, a foot either way will correspond to less than one-quarter of a dioptre of inexactness. Hence, for accurate work it is best to make the determination of the point of reversal at the greatest distance at which it can be certainly made in the visual zone. The importance of this has been especially dwelt upon by Randall (*Trans. Section on Opthalmol. Am. Med. Assoc.*, 1894, p. 63).

DISTANCE OF THE SURGEON FROM THE PATIENT.

What this distance may be will vary in different eyes. In general, it is limited by the size of the area in which the movement of light and shade is to be watched. The pupil fully dilated may be eight or ten millimetres in diameter, and movement across the whole width of such a pupil could be readily watched at a distance of 4 to 6 metres. But the diameter of the visual zone of the pupil, the only area in which the movement is of practical importance, is commonly much less than this, say from 4 to 6 millimetres; and the movement of light across it can only be satisfactorily studied within the distance of two or three metres. Beyond one metre, however, the necessary inaccuracies of distance become usually of slight practical importance.

In cases of aberration invading the central portions of the pupil and still more in cases of irregular astigmatism, the visual zone is considerably less in area than in the ordinary normal eye. In these cases, the test must be applied from a still shorter distance, often one-half or one-third of a metre, or even less.

With the plane mirror it is easy to adopt any distance that suits the particular case. With the concave mirror any considerable variation in the distance requires a corresponding variation in the focus of the mirror used; a mirror of shorter focus being employed when the distance between the observer and patient must be short; and of longer focus if a greater distance is to be maintained.

The reason for this is that if the concave mirror be brought too close to the observed eye it gives an immediate source of light relatively too large and too far in front of of the observer's position when at the point of reversal; while if it be removed too far from the patient's eye, the diffusion of the light over a larger area is so rapid that it gives an illumination that is too feeble. These changes are much more rapid with the concave than with the plane mirror; as one may readily demonstrate by holding both mirrors in his hand in the darkened room and reflecting areas of light upon a wall from various distances. I have elsewhere (*Journal of the Am. Med. Assoc.*, Sept. 4, 1886) demonstrated the relations of the one to the other. In general the distance at which a concave mirror can be used to best advantage is a little over four times its focal distance.

For the majority of cases then, a distance of from ½ to 2 metres is convenient for the plane mirror; and one metre or a little less for the concave mirror, having the usual focal distance of from 20 to 25 centimetres.

Final Test.—When it is desired to make the shadow test as accurate as possible, it is well to complete the test by placing before each eye lenses representing its supposed correction, with such addition to the convex or diminution of the concave spherical as shall bring the point of reversal to the greatest distance at which the movements of light and shadow can be satisfactorily studied in the particular eyes in question; and from that distance to test the movement of light and shade, looking especially for uncorrected astigmatism, and comparing the one eye with the other for any evidence of remaining inequality of refraction.

CHAPTER IV.

REGULAR ASTIGMATISM.

The essential fact of regular astigmatism is that in two different directions, at right angles to each other [the principal meridians] the curvature of the dioptric surfaces differs, so that they exert unequal refractive power; and that in all other directions, or meridians, the refractive power bears such a relation to the refractive power of these principal meridians, that it is only necessary to consider what happens in their direction.

Two Points of Reversal.—In such an eye, the rays coming from the same point of the retina, and passing out through surfaces that refract unequally in different meridians, must leave the eye with different degrees of divergence, or convergence, in the directions of these different meridians. If after passing out the rays are convergent, or rendered so by passing through a convex spherical lens, they will be more convergent in one principal meridian than the other, and the point of reversal for one principal meridian will be nearer the eye, than the point of reversal for the other principal meridian. The position of each point of reversal gives the amount of myopia (either original or produced) in the principal meridian to which it belongs. The difference between the amounts of myopia in the two principal meridians is the amount of astigmatism. The general plan of measuring astigmatism by skiascopy is to ascertain the point of reversal and measure the degree of myopia for each principal meridian separately; and then, by subtracting the one amount from the other, to find the amount of regular astigmatism.

The Band-like Appearance.—This difference in the position of the points of reversal for the different meridians, gives rise to certain phenomena of great practical importance in skiascopy. It is true of the astigmatic as of the non-astigmatic eye, that, as the point of reversal is approached, the image of the retina seen through the pupil becomes magnified (see page 29). And, it necessarily follows that when the observer's eye is nearer to the point of reversal for one meridian than it is to the point of reversal for the other meridian, the retinal image is more magnified in the direction of the principal meridian, to which the nearer point of reversal belongs.

When the observer's eye is placed at the point of reversal for one meridian, the retinal image becomes indefinitely magnified in the direction of that meridian, while comparatively little magnified in the direction at right angles to it. Each point of the retina then appears in the pupil as a line running in the direction of that principal meridian, and the retinal light area, which consists of a number of these points, takes the form of an elongated band of light, running in the direction of the principal meridian which has its point of reversal at the observer's eye. This is the *band-like* appearance of the light in the pupil, characteristic of astigmatism; and the less illuminated part of the pupil beside it is the "linear shadow" of Bowman. Figure 7 represents this appearance when the eye is placed at the point of reversal for one principal meridian, represented about twenty degrees from the vertical; and figure 8 represents the appearance presented at the point of reversal for the other principal meridian, twenty degrees from the horizontal.

Its *direction* is always that of the principal meridian, at whose point of reversal it is seen; and it is more pronounced:

In proportion to the degree of astigmatism:

The nearness of the observer's approach to the point of reversal:

And the perfection of the focusing of the light upon the retina in the direction perpendicular to this principal meridian, that is, in the other principal meridian.

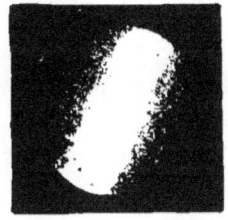 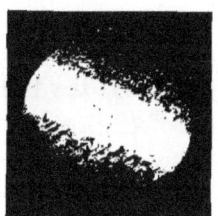

Fig. 7.   Fig. 8.

In estimating astigmatism by skiascopy, two distinct things are to be done, which require different arrangements of the source of light. The first is to determine accurately the direction of the principal meridians by bringing out most distinctly this band-like appearance in the pupil, indicating the direction of one of these principal meridians; the other being always, for regular astigmatism, at right angles thereto. The second thing to be done is to measure accurately the refraction in each of these principal meridians.

The test proceeds at first as for myopia or hyperopia in a non-astigmatic eye, until a point of reversal is found. Then it is discovered that this point of reversal is only for the movement of light and shadow in one direction, and does not hold for movements at right angles to that direction. The observer has now brought his eye to one point of reversal where the band-like appearance can be best perceived. But, as he has been working with the original source of light in the position most favorable for the measurement of hyperopia and myopia, the position that brings the immediate source of light as close as possible to the mirror (see page 37), he will probably see very little appearance of the band in the pupil, even with the higher degrees of astigmatism. The reason for this is, that with the immediate source of light in this position, the light is most accurately focused on the retina in the direction that the band should take. And, in the direction at right angles to the band, the focusing is quite incomplete, so that the diffusion at what should be the sides of the band partly or entirely neutralizes the effect produced by the greater magnification of the retina in the direction of the band, which, otherwise, would cause the band-like appearance.

In order to bring out this band-like appearance, it is necessary to make the focusing from side to side of the band as perfect as possible. And, to secure the perfect focusing in the principal meridian at right angles to the one in which the band is sought, the immediate source of light must be brought to the point of reversal for that other principal meridian. *The band-like appearance is most perfectly developed when the observer's eye is at the point of reversal for one principal meridian, and the immediate source of light at the point of reversal for the other principal meridian.*

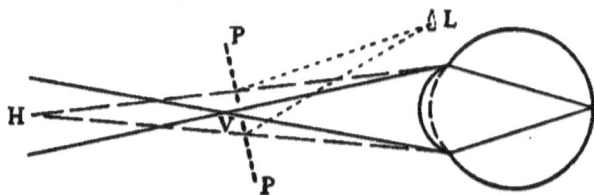

FIG. 9.

In figure 9, the solid lines represent the vertical meridian of an astigmatic eye; and the rays emerging, so turned in that meridian as to give the point of reversal at V. The broken lines represent the less curved horizontal meridian of the cornea, and the rays so turned in that meridian as to give a point of reversal at H. The dotted lines represent a plane mirror, PP, with the eye of the observer at V, and the light L pushed off from the mirror, so that the rays enter the eye as though they came from H, and are perfectly focused on the retina in the horizontal meridian, rendering most distinct the appearance of a vertical band.

For illustration, suppose a case [which the student will do well to reproduce for actual study, either in the artificial eye or by lenses placed before the living eye] having compound myopic astigmatism, the vertical meridian of the cornea being 2 D. myopic and the horizontal meridian 1 D. myopic. When, with the plane mirror, the observer's eye is one-half metre from the observed eye, it will be at the point of reversal for the vertical meridian, and in a position to see a vertical band of light. But, if the source of light be placed as close to the mirror as possible, the rays from it will be the more accurately focused upon the retina in the vertical meridian and more diffused horizontally, so that the *real* form of the retinal light area will be rather that of a horizontal line or band.

Now, from the observer's position, the retina is most magnified in the vertical direction, and this vertical magnification would cause a point of light on the retina to *appear* as a vertical band in the pupil; but, with the light area really in the form of a horizontal band, the effect of the vertical magnification is largely neutralized, and the appearance in the pupil may be quite indefinite.

To bring out the band-like appearance: While keeping the observer's eye and mirror in the same position, the original source of light must be pushed off from the mirror one-half metre, the immediate source then retreats correspondingly behind the mirror, and approaches the position of the point of reversal in the horizontal meridian, one metre from the eye.

With the light and mirror in this relation to the eye, the rays are perfectly focused upon the retina in the horizontal meridian and diffused in the vertical meridian, so that the real form of the retinal area of light is a vertical line or band. This vertical line or band being viewed from the point of reversal of the vertical meridian (where it will be greatly magnified in the vertical direction and but slightly magnified in the horizontal direction), gives rise to the appearance of the most distinct vertical band of light in the pupil. And, under these conditions, the presence of the astigmatism and the direction of one of its principal meridians is most clearly and accurately revealed.

Taking the same case and using the concave mirror at a distance of one metre, which is the point of reversal for the horizontal meridian, the appearance of a horizontal band of light in the pupil may be rendered most distinctly visible. But, in order to develop it clearly, it will be needful to bring the original source of light to such a position that the immediate source will be one-half metre in front of the mirror; that is, one-half metre in front of the observed eye, at the point of reversal for the vertical meridian. For it is from this position the light will be most perfectly focused on the retina in the vertical meridian, while diffused in the horizontal meridian, and the greater horizontal magnification of the retina at the point of reversal for the horizontal meridian where the observer's eye is placed, will emphasize and increase the appearance of the horizontal band of light then thrown on the retina.

Since, with the plane mirror, the immediate source of light is always back of the mirror, and cannot be brought in front of it, with it the direction of the band can only be accurately determined for the meridian whose point of reversal is nearest the eye. It is only with the eye and mirror at this point of reversal that one is able, with the plane mirror, to bring the immediate source of light to the other point of reversal. And, with the concave mirror, since the immediate source of light is always in front of the mirror, the band-like appearance can only be distinctly brought out in the meridian which has its point of reversal the farther from the eye, as only with the eye at that point of reversal can the immediate source of light, with the concave mirror, be brought to the other point of reversal.

With either the plane or concave mirror, only the band in one of the principal meridians can be most distinctly developed. But it is unnecessary in practice to bring out the bands in both meridians, since, by knowing the direction of one principal meridian, the other being always perpendicular to it, is also known.

The measurement of the refraction in either of the principal meridians of astigmatism, is quite similar to the measurement of refraction in hyperopia and myopia. To determine whether the movement of light in the pupil in a certain meridian is with or against the movement of light upon the face, it is necessary that the focusing of the light on the retina be as perfect as possible in that particular meridian. To secure this, the immediate source of light must be as close as possible to the position of the observer's eye. Hence, having determined the existence of the astigmatism and the direction of its principal meridians, the measurement in these meridians will proceed as the measurement of myopia or hyperopia.

Changes in the Light Area at Different Distances.— In regular astigmatism, supposing the eye to be myopic in all meridians, or a convex lens placed before it sufficiently strong to over-correct the hyperopia in all meridians, the observer using a plane mirror and viewing the eye from different distances, will be able to recognize the following changes in the appearance and movement of the light in the pupil:

From a position within the point of reversal of the more myopic meridian, the light will be seen to move *with* the light on the face, in all directions. As the observer's eye is withdrawn from the observed eye, and approaches the point of reversal for the more myopic meridian, the light area in the pupil becomes elongated in this meridian; and, while the movement is still with the light on the face in all meridians, it becomes more rapid in the direction of this elongation than in the direction perpendicular thereto.

The observer withdrawing his eye still farther, on reaching the point of reversal for the more myopic meridian, [V, in figure 9,] is unable to distinguish the movement in this meridian, while the movement in the meridian at right angles to it is still *with* that of the light on the face.

This point being reached, if the original source of light be pushed away from the mirror, so that its reflection, the immediate source of light approaches the point of reversal for the less myopic meridian, the form of the light in the pupil becomes a distinct band running in the direction of the more myopic meridian, readily seen to move from side to side, but without perceptible movement in the direction of its length.

Bringing the source of light back to its usual position close to the mirror, and withdrawing his eye still farther from the eye under observation, the observer again sees the movement of the light in the pupil in all directions. But in the direction of the most myopic meridian, it is now *against* the light on the face; while in the meridian at right angles to this, it is still *with* the light on the face. The band-like appearance is now lost entirely; the area of light in the pupil taking at one distance the same shape as though no regular astigmatism were present.

But, as the point of reversal for the less myopic meridian is approached, elongation in the direction of that meridian may be noticed, and the movement of the light in that meridian *with* the light on the face becomes more rapid than the movement *against* the light on the face now seen in the more myopic meridian. When the point of reversal for the less myopic meridian [*H*, figure 9] is reached, the movement in its direction ceases, but it is impossible, at this point (with the plane mirror), to bring out so distinct a band as was seen in the direction of the other meridian.

Withdrawing still farther, the light in the direction of the less myopic meridian begins to move *against* the light on the face, at first very rapidly as compared with the movement in the more myopic meridian. But, as the observer withdraws farther from this second point of reversal, the difference in rate of movement in the two meridians becomes less noticeable.

With the concave mirror, the same series of appearances are presented, except that the directions of movement are reversed—the erect image seen from within the point of reversal giving movement of the light in the pupil *against* the movement of the light on the face, and against the mirror; and the inverted image seen from beyond the point of reversal giving movement of the light in the pupil *with* the mirror and with the light on the face. (See page 26.) With the concave mirror the meridian in which it is possible to bring out the band-like appearance of the light most distinctly is the meridian of less myopia. With such a mirror it will also be necessary to bring about the series of changes in the movement of the light area, which has been referred to, by changes of the lens placed before the eye, and not by changes in the observer's distance from the eye studied.

Direction and Movement of the Bands in Astigmatism.—The reason for the constant conformity of the direction of these bands of light to the principal meridians of refraction is obvious from their dependence on the magnification of the retina. That conformity sharply separates them from the somewhat similar appearance seen near the point of reversal in eyes free from astigmatism (page 31).

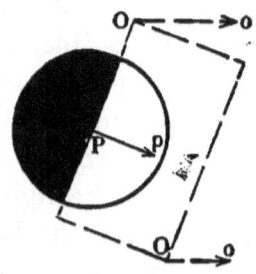

Fig. 10.

The apparent movement always at right angles to their direction is dependent on an optical illusion, of which one may satisfy himself by making a hole in the centre of a sheet of paper, holding behind this hole the edge of a card, and moving it in a direction oblique to this edge. The motion will appear to be in a direction nearly or quite perpendicular to the edge seen.

Thus, in figure 10, the real movement of the card behind the opening, or the band of light behind the pupil, may be in the direction $O\ o$. But the movement will appear to be in the direction $P\ p$.

CHAPTER V.

ABERRATION AND IRREGULAR ASTIGMATISM.

In astigmatism, strictly regular, though the refraction differs in different meridians, in any given direction or meridian it is the same at all parts of the pupil. In aberration and irregular astigmatism, the refraction differs in different parts of the pupil, even in the same meridian. All eyes present variations of this kind; and these variations constitute an obstacle to the measurement of refraction, by skiascopy or by any other method.

Appearances of Irregular Astigmatism.—To the beginner with skiascopy, these constitute the most serious obstacle he has to encounter. By one who has thoroughly mastered the principles of the test and become familiar with the various appearances of light and shade in the pupil, mistakes due to aberration or irregular astigmatism are readily avoided; and to the experienced skiascopist is revealed the reason for uncertainty in the results obtained by other methods, or of failure to secure perfect vision on account of these defects.

If we suppose two parts of the pupil, one of which has its point of reversal at the observer's eye, while the other is at a considerable distance therefrom, the illumination of the former will be the more feeble, of the latter the more brilliant; the movement of the light in the former, if perceptible, will be rapid, in the latter, slow. If one watches two parts of the pupil, one of which has its point of reversal back of the observer's eye, and the other in front of it; in the former the light will have a direct and in the latter an inverted movement.

With the irregular astigmatism due to preceding corneal inflammation, or to the changes in the refraction of the lens that sometimes precede cortical cataract, the pupil appears broken up into a considerable number of distinct areas, each of which has its separate movement of light and shadow, constituting the typical ophthalmoscopic or skiascopic picture of irregular astigmatism. The appearance

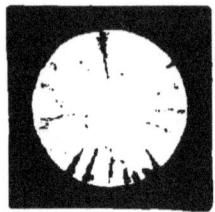

FIG. 11.   FIG. 12.

caused by irregular astigmatism following corneal disease is shown in figure 11. That due to changes in the lens such as may precede cortical senile cataract is shown in figure 12, in which the black lines represent fixed spicules of actual opacity, while the other parts of the pupil indicate merely refractive differences, and change from light to dark, or dark to light, as the inclination of the mirror is varied. Some such appearance is sometimes presented by young persons, indicating a congenital defect which may not noticeably increase in many years.

If the differences of refraction in the different parts of the pupil are slight—that is, if the aberration or irregular astigmatism is of low degree—these differences of illumination and movement will not be perceptible until the observer brings his eye close to the point of reversal. But at the point of reversal, they become perceptible and constitute a striking phenomenon in almost all eyes; and, to the observer who does not understand their significance, one that is extremely confusing. In the nature and arrangement of its irregular astigmatism, every eye is peculiar. The varieties of play of light and shade that are obtainable as the point of reversal is reached, are as numerous as the eyes examined. Even the two eyes of the same individual differ.

The only practical way to deal with irregular astigmatism by skiascopy is to understand thoroughly the general optical principles of the test, and apply them, so far as may be needful, in the individual case. Certain peculiar forms of variation of the refraction of the eye in different parts of the pupil are, however, of sufficient constancy, regularity and practical importance, to warrant their separate classification and study. The most important of these is the regular, or symmetrical, aberration of the eye.

Symmetrical Aberration.[1]—This is an error of the refraction of the eye which causes the rays of an incident pencil falling on the same meridian of the cornea, but at different distances from the axial ray, to meet at different distances behind the cornea, while rays piercing different meridians of the cornea, at the same distance from the axial ray, intersect it at the same point. It is a defect similar to the monochromatic aberration of convex and concave spherical lenses. It is readily recognizable in almost all eyes by skiascopy. In the majority of cases, it is in the same direction as ordinary spherical aberration; that is, near the margin of the pupil there is a stronger lens action than near the centre—the rays entering through the margin are brought to a focus first, the rays entering nearer the centre being focused farther back. This is called *positive aberration*.

In a certain proportion of cases, however, the defect is in the opposite direction; the rays passing near the centre of the pupil being brought first to a focus, and those pass-

---

[1] For an account of this error of refraction see paper by the author, *Trans. Amer. Ophthalmological Society*, 1888, p. 141.

ing through the periphery being focused farther back. The centre of the pupil has the stronger, and the periphery of the pupil the weaker, lens action. This is *negative aberration*.

The Visual Zone.—The variation of refraction, however, does not usually proceed regularly from the centre of the pupil to the margin. But, as with spherical lenses, and to a greater degree, the central refraction is comparatively uniform over a considerable area; and, towards the margin the change of refraction becomes progressively more marked. This area in the centre of the pupil of comparatively uniform refraction is the usual *visual zone*. It is the portion of the pupil that is of practical importance for purposes of distinct vision. Its size varies considerably. Sometimes it includes almost the whole of the dilated pupil, in other eyes an extremely small area near the centre of the pupil will be regular, and the remainder of the pupil useless for accurate vision. If a high degree of irregular astigmatism be present, the visual zone, instead of being a central area of considerable size, will often be some particular portion of the undilated pupil, which happens to have the most regular curvature.

In any case, for the correction of ametropia, it is the behavior of the light and shade in the visual zone which has to be studied. Its behavior elsewhere may be disregarded. It is often much easier to watch the movement of light and shade in some other portion of the pupil—some part of the *extra-visual zone.* And, if the observer does not understand their relative importance, he will be apt to fix his attention on this latter, and be led away from the true refraction of the eye he is examining. This is the more likely to happen, because in that part of the pupil, which has its point of reversal at, or near, the observer's eye, the direction of the movement of light and shade is difficult to see, while in other portions, the movement is more striking.

The Appearances of Positive Aberration.

The Appearances of Positive Aberration.—The appearances presented by an ordinary case with positive aberration may be considered in the order in which they will be developed with the *plane mirror*, the observer starting to examine the eye from within the point of reversal for the most myopic part of the pupil, and gradually withdrawing his eye until it is beyond the point of reversal for the least myopic part of the pupil. From the first position, the light area in the pupil is seen to move with the light on the face entirely across the pupil; its motion in the edges of the pupil being more rapid than in the centre. If, now, the observer's eye is withdrawn to the point of reversal for the margin of the pupil, there appear in the margin points in which no movement of the light can be seen. Some of these may be points of stationary light, and others, points of stationary shadow.

As the observer's eye is still farther withdrawn, the points of stationary light run together and form a complete ring of light in the periphery of the pupil, shown in figure 13, which is presently seen to have an inverted motion—to move against the light on the face. Within this is a ring

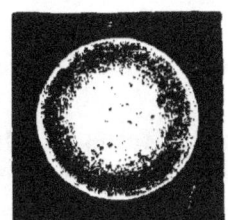

FIG. 13.

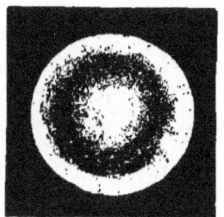

FIG. 14.

of comparative shadow where the movement is swift, and difficult or impossible to recognize; and still within this lies an area of light—what remains of the light area first seen, now considerably reduced in size—which still moves with the light on the face.

As the observer draws still farther back, this area of light at the centre of the pupil, as shown in figure 14, grows smaller, and its movement more difficult to certainly distinguish. The ring of comparative shadow around it encroaches upon it, and the ring of light in the margin of the pupil in turn encroaches upon the shadow, and becomes brighter and its movement more readily noticeable.

Withdrawing still farther, the point of reversal for the centre of the pupil is reached. The central area of light becomes faint and its movement ceases to be noticeable, the ring of feeble illumination surrounding it having swallowed it up. But, around this feeble light area, the ring of inverted movement has now grown broad and distinct. And, as the observer withdraws still farther, this ring of inverted movement closes in until it occupies the whole of the central area, and the observer sees an area of light moving across the whole pupil, having an inverted movement—that is, against the mirror or light on the face.

The movements of these erect and inverted light areas in the pupil are illustrated by figures 15 and 16. Figure 15 shows the plane mirror turned to the left, or the concave mirror turned to the right, the central erect area being displaced toward the left, and the peripheral inverted area toward the right of the space it occupies. Figure 16 represents the light areas displaced in the opposite directions by an opposite inclination of the mirror.

With the *concave mirror*, a similar series of changes may be brought about by placing before the eye successive strengths of the lenses, beginning with the weakest convex or strongest concave. The first should allow the points of reversal for all parts of the pupil to be back of the observer, and the successive changes bring these points closer and closer to the observed eye until all are in front of the observer. The movement is at first *against* the light on the face. Then appears the ring of illumination and swift movement in the margin of the pupil *with* the light on the face. The central area of light is then encroached upon by the ring of faint illumination, and this in turn by

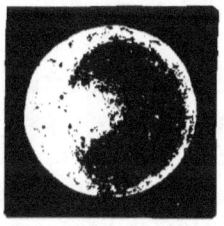
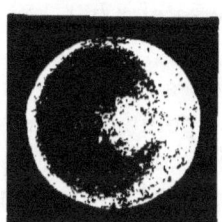

FIG. 15.   FIG. 16.

a ring of more brilliant illumination in the margin, moving with the light on the face, which latter finally occupies the whole area of the pupil.

If the point of reversal be approached from the opposite direction, that is, starting with the observer's eye beyond it, we have, with the *plane mirror*, at first, inverted movement across the whole pupil. Then, as the point of reversal for the centre of the pupil is approached, the light in the central zone becomes feeble and its movement indefinite. When the point of reversal for that part is passed, there appears, in this central zone, an erect movement of light and shade, at first rapid and hard to see, but growing slower, gaining in distinctness, and occupying a larger and larger part of the pupil as the patient's eye is approached, until, finally, it occupies the whole area.

With the *concave mirror*, starting with the strongest convex or weakest concave lens, the movement is first with the mirror throughout the pupil, then, as the lens is changed, it becomes indefinite at the centre; presently it is against that of the mirror at the centre, while still with it at the margin; and, with still weaker convex lenses, or stronger concaves, it becomes against the mirror throughout the whole pupillary area.

Appearances of Negative Aberration.—With negative aberration, the series of changes is apt to be less regular and complete, and the picture presented by the pupil is less characteristic. But the succession of appearances is the reverse of what has been described for positive aberration.

With a *plane mirror* starting closer to the patient's eye than the point of reversal for the most myopic part of the pupil, the movement is with that of the light on the face throughout the whole pupil. As the observer's eye is withdrawn to a greater distance, this movement becomes indefinite, and the light feeble near the centre of the pupil. Presently, the movement at the centre of the pupil is lost, while still quite distinctly with that of the light on the face in the irregular ring-shaped area of the periphery. Withdrawing still farther from the eye, the inverted movement at the centre of the pupil becomes distinctly visible, and the direct movement near the margin becomes more and more encroached upon and less and less distinct, until finally all erect movement is lost and we have only the inverted movement, which extends across the whole pupil. Before the erect movement entirely disappears, it is apt to break up into small areas detached from one another by spaces of comparative shadow, but each presenting some remnant of the erect movement.

With the *concave mirror*, starting with a convex lens so weak, or a concave lens so strong, that the point of reversal is back of the observer, we have the movement against the light on the face throughout the pupil. The strengthening of the convex lenses or the weakening of the concaves, so as to bring the point of reversal close to the observer's eye: causes, first, the fading of the light and the indefiniteness of its movement in the centre of the pupil, then the movement, with the light on the face, at the centre, and the area of this movement extending until it includes the whole of the pupil.

Approaching the point of reversal from beyond it, we have, with the *plane mirror*, inverted movement throughout the whole pupil, giving place to indistinctness first at the margin. Then the indirect movement confined to the central area of the pupil and direct movement appearing at certain parts of the marginal area. This direct movement becomes more distinct and its area increases as the patient's eye is approached, until, at the point of reversal for the centre of the pupil, all inverted movement is lost, and the erect movement is seen in all parts of the pupillary area.

With the *concave mirror*, starting with the point of reversal between the observer and patient, and removing it successively farther from the patient, by the use of weaker convex or stronger concave lenses, we have first the movement with the light on the face throughout the pupil, then indefinite at the pupillary margin, changing, in turn, to movement against the light on the face. The area of this peripheral movement then encroaches upon the central area until that is obliterated, and the movement against the light on the face occupies the whole width of the pupil.

While the order of their development remains the same, the exact character of the appearances presented varies considerably with the degree of aberration. Generally, in the higher degrees, the areas of light occupy the greater part of the pupil and the area of feeble illumination separating them is comparatively narrow. While in very low degrees of aberration, the area of feeble illumination is broad, and it may be difficult to recognize more than one of the light areas at one time. That is, when the area of erect movement is visible, the remainder of the pupil is occupied by the area of feeble illumination; and when the area of inverted movement is developed, the area of feeble illumination so encroaches upon the area of direct movement that the latter can no longer be identified.

In some eyes, the variation of refraction from point to point, which constitutes symmetrical aberration, is almost or entirely confined to the periphery of the pupil. In these eyes the appearances characteristic of aberration are hard to develop.

Appearance of Conical Cornea.—In other eyes, an opposite condition is present. The variations of refraction, instead of being confined to the periphery of the pupil, encroach upon the normal visual zone, confining it to a very narrow area. In these eyes, the skiascopic appearances of aberration are striking and characteristic, and one of them is that which has been regarded as peculiar to conical cornea.

The error of refraction produced by conical cornea is a high degree of negative aberration. At the apex of the cone, the curve is sharp, causing, usually, very high myopia in the corresponding part of the pupil. The sides of the cone, on the other hand, are comparatively flat, causing diminished myopia as the region of the apex is departed from and sometimes running into hyperopia near the edge of the pupil.

If the observer's eye be placed somewhere near the point of reversal for the periphery of the pupil, the movement of light in that portion of the pupil will be rapid, but the movement in the portion of the pupil corresponding to the apex of the cone will be slow. On account of the high myopia, the point of reversal for this part of the pupil is very close to the eye, and, generally, many dioptres removed from the observer's eye. The movement of light in the pupil, then, is slow near the centre and rapid towards the periphery, causing the area of light to appear to wheel around a fixed point corresponding to the apex of the cone. The light area is first seen on one side of the pupil, then on the other, but always rests upon the central fixed point.

In certain positions of the light, the form of this area

will be somewhat triangular, its base resting on the margin of the pupil and its apex at the apex of the corneal cone. Sometimes the triangle covers almost half of the pupil, in other conditions of light it is considerably narrower, but the constant and characteristic phenomena is the wheeling of the light area about the fixed point at the apex.

This is shown in figures 17 and 18, which represent the appearances of the pupil with the mirror inclined in opposite directions.

It was for the detection of these appearances, to which attention was called by Bowman, in 1857, that the test was first employed. Bowman mentions that he was able by means of it to detect low degrees of conical cornea, which could not be detected in any other way. It is certain that among those cases that have been classed as low degrees of conical cornea, on account of their presenting such appear-

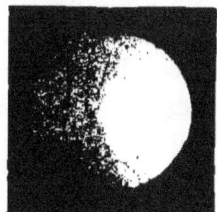
FIG. 17.

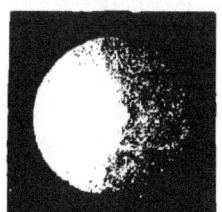
FIG. 18.

ances, a considerable proportion were not of conical cornea at all, but were cases of high aberration from other forms of defect in the dioptric surfaces.

The appearances in question occur in all cases of high aberration. Where the aberration invades the central portion of the pupil, and is not confined to the periphery, the phenomena are quite as striking and characteristic, as in cases of conical cornea. And cases of high positive aberration are more common than cases of true conical cornea. The conditions for their recognition are that the observer's eye shall be comparatively near the point of reversal for the margin of the pupil, and comparatively far removed (estimating by dioptres) from the point of reversal for the centre of the pupil. By careful management of the light and relative position of observer and patient, something of such an appearance can be demonstrated in the majority of eyes.

Like the band-like appearances of the light in astigmatism, those of conical cornea reveal the presence of the condition and the location of the apex of the cone, but beyond this, they are of little value. The measurement of the difference of refraction between the margin and the centre of the pupil, or the measurement of the refraction in the portion of the pupil best suited to purposes of vision, must be accomplished by the same application of skiascopy as serves to measure the amount of hyperopia or myopia in an eye free from astigmatism and aberration.

The series of movements presented in positive aberration can be well studied in one of the numerous forms of artificial eye, in which spherical lenses are used to represent the dioptric surfaces ; and it is well by such study to become thoroughly familiar with them. They may, of course, be studied in living eyes presenting positive aberration ; but, in many of these, the appearances are not so typical, and regular in order of sequence, as with the ordinary strong spherical lenses. The appearances presented by negative aberration can only be studied in eyes in which this condition of the refraction is present, but their recognition and observation will be comparatively easy to one who has mastered the corresponding appearances of positive aberration, and who understands the optical conditions on which they depend.

Scissors-like Movement.—A special form of irregular astigmatism exists, of sufficiently frequent occurrence and striking character to merit special description. It is also of some practical importance. In it, one portion of the pupil, as an upper or lower half, is more myopic in a certain meridian than is the other part of the pupil. This causes an inverted movement of light in the one portion of the pupil, while there is an erect movement in the other. These two areas are distinct, and separated by an intermediate zone of feeble illumination. As the light is made to move back and forth in the proper direction, the two areas of light in the pupil are seen alternately to approach and separate, narrowing or widening the intermediate zone. As the areas, under these circumstances, are generally band-like, or have comparatively straight margins, the effect is similar to that of the opening and closing of a pair of scissors. These appearances are represented in figure 19, which shows them with the mirror so turned as to separate the two areas; and figure 20, which represents them brought together by an opposite inclination of the mirror. Supposing the upper part of the pupil to be more myopic, figure 19 corresponds to the plane mirror facing down or the concave mirror facing up; and figure 20 shows the plane mirror facing up or the concave mirror facing down.

FIG. 19.

FIG. 20.

The relative size of the two areas will depend on the distance of the observer from the eye or upon the strength of the lens employed. As the observer withdraws to a greater distance, or the convex lens is made stronger, or the concave lens is made weaker, the area of inverted movement encroaches upon the zone of feeble illumination separating the areas of light and the area of erect movement diminishes. As the observer comes closer to the eye, or the convex lens is made weaker or the concave lens stronger, the area of inverted movement diminishes. Always the observer's eye is at or near the point of reversal for the portion of the pupil occupied by the intermediate zone of feeble illumination; and, in making the determination of the refraction for practical purposes, care must be taken to see that this zone occupies a portion of the pupil that is available when the pupil is contracted, as under ordinary conditions of illumination and near work.

The scissors-like movement may be produced in an artificial eye by placing the lens which represents the dioptric surfaces, so that the light passes through it obliquely. It may also be developed in most eyes by applying skiascopy in some direction at a considerable angle to the optic axis. Its presence in the eye indicates obliquity or imperfact centreing of one or more of the dioptric surfaces. Probably it is often due to some obliquity in the position of the crystalline lens. Perhaps, because of such obliquity, this appearance of light and shadow in the pupil is apt to co-exist with a considerable degree of regular astigmatism, which, on account of it, becomes more difficult to recognize and measure than it would otherwise be. Eyes presenting it, therefore, demand special care and attention on the part of the observer, to develop the best vision they are capable of with correcting lenses.

CHAPTER VI.

PRACTICAL APPLICATION WITH THE PLANE MIRROR.

Position and Arrangement of Light.—The room being thoroughly darkened, the patient and surgeon take positions facing each other at a distance of about one metre, with the original source of light close to the surgeon on the side of the eye he desires to use, that is on the right if he intends using his right eye for the test. He can really see the movement of light and shade in the pupil with but one eye at a time; yet he will find it more pleasant to work with both eyes open if he once learns to do so. The source of light should be freely movable from fifteen centimetres in front of the patient's face to over a metre away, a movement obtainable with a double jointed bracket of over one-half metre total length. The light is covered from the

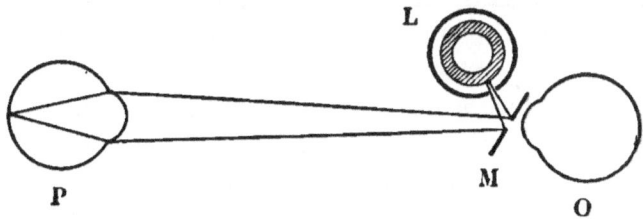

FIG. 21.

patient's face, and also from the surgeon's except at the aperture of about five millimetres opposite the brightest part of the flame. The arrangement of the surgeon and patient with reference to the light is shown in Fig. 21, in which L represents the light, M the mirror, and O and P the eyes of the observer and patient.

The mirror is held so that with the eye behind it the surgeon can watch, through the sight hole, the movement of light on the patient's face; and turned until the area of light that it reflects falls upon the eye to be tested. If difficulty is experienced in properly directing the light, the surgeon may hold his hand a few inches in front of the mirror and upon it find the light area and get it properly directed towards the patient's eye. If the mirror be large it is necessary that the central portion of the light area be made to fall upon the patient's eye, the centre being marked by the spot of feeble illumination, corresponding to the sight hole of the mirror.

With the light properly directed, the pupil appears to be partly or wholly occupied by a red glare, the light area with which skiascopy is especially concerned. In first attempting the test, care must be taken to discriminate clearly between this general red glare and the reflection from the cornea or from the surfaces of any lens that may be placed before the eye. These reflections have the same color as the light used for the test. The reflection from the cornea is small and brilliant, a mere point of light, if the room be thoroughly darkened and the original source of light properly shaded. The reflections from the lens employed are larger and more confusing. They may be avoided by tilting the lens slightly, which causes them to pass off to the periphery, leaving the centre of the lens free from reflection.

Hyperopia.—If the mirror be rotated about a vertical axis, that is if it be made to turn more to the right or left, the area of light in the pupil will be seen to move *with* the light on the face to the right or left as the inclination of the mirror changes. If the rate of movement be slow, the hyperopia is of high degree, if more rapid, it is lower.

A convex lens is now to be placed before the eye, and this rate of movement of light in the pupil is the guide to the probable strength of lens required. If the observer has not sufficient experience with skiascopy to judge in this way about the strength of the lens required, he will save time by placing before the eye rather a strong lens, one of say 5 D. With this the light is again thrown upon the eye, and if the lens be not sufficient to correct the hyperopia present, the movement of light in the pupil will still be *with* that of the light on the face. In this case a still stronger lens must be used. This strengthening of the convex lens before the eye is continued until one is found which causes the reversal of the apparent movement of light in the pupil—until the light in the pupil moves *against* the light on the face.

Then the surgeon is to approach the patient, meanwhile rotating the mirror and watching for the nearest point at which he still sees the inverted movement in the visual zone. Near this point, the illumination of the pupil becomes quite feeble; and the movement, being rapid, requires the closest watching. Approaching still nearer to the patient the light in the visual zone is seen to move *with* the light on the face, and the greatest distance at which this can be distinguished is to be noted. Between these two, the least distance of inverted movement, and the greatest distance of direct movement lies the point of reversal. But, for reasons given page 39, it is better to take the latter, the greatest distance at which direct movement can be perceived as the point sought.

The distance from the surgeon's eye to the patient's eye is then measured. It is the focal distance of the amount of myopia produced by the convex lens employed. That amount is to be subtracted from the total strength of the lens to ascertain the proportion of its strength which has been necessary to correct the existing hyperopia.

Having made such a determination of the refraction and having repeated the various observations until no doubt is left as to their correctness, the lens before the eye is to be changed for one sufficiently weaker to carry the point of reversal to as great a distance, as the size of the visual zone will allow for the accurate determination of the movements of light and shade within it. At this distance the estimate of the ametropia is to be completed.

For example: Suppose the eye under examination to have hyperopia of 3 D. When the 5 D. lens is placed before it, the point of reversal will be brought to one-half metre. As the surgeon's eye is made to approach that of the patient, the inverted movement in the visual zone will cease when they are about 60 centimetres apart. Going still closer, the erect movement will be distinguished at about 40 centimetres. These observations are to be repeated until the surgeon makes sure that the point of reversal lies somewhere between 40 and 60 centimetres. The 5 D. lens is then replaced by the 4 D. lens. Repeating the test, the inverted movement is seen as near the eye as one and one-quarter metres and the direct movement almost as far away as one metre. This locates the point of reversal at about 1 metre from the eye, and determines the myopia caused by a 4 D. convex lens to be 1 D., and the refraction of the eye to be 4. D.$-$1 D.$=$3 D. of hyperopia, with less than 0.25 D. of possible error either way.

Myopia.—In myopia the first rotation of the mirror will usually cause a movement of light in the pupil *against* that of the light on the face. The surgeon then approaches the patient, continuing the movement of the mirror and watching the apparent movement of light in the pupil, until this apparent movement becomes rapid and indefinite and presently is entirely lost. Approaching still closer to the patient's eye, the movement of the light area in the pupil again becomes distinct, but is now *with* the movement of the light on the face. Drawing back again, the surgeon notes the greatest distance at which this erect movement can be observed, and then again the shortest

MYOPIA.

distance at which the inverted movement is distinguishable, and takes a point midway between these to be the point of reversal.

The distance of this point of reversal from the patient's eye is the focal distance of the lens that will be required to correct the myopia. To complete the test, however, a lens about 1 D. weaker than this is placed before the eye to bring the point of reversal to the distance of one metre; and the test is repeated, the surgeon noting carefully the greatest distance at which the erect movement is visible, and the shortest distance at which the inverted movement is perceived, always in the visual zone. The distance of the point of reversal as thus determined is the focal distance of the lens required to correct the remaining myopia. The strength of such a lens, added to the strength of the lens already before the eye, gives the total amount of myopia present.

Suppose the eye to be 6.5 D. myopic. With the first test the inverted movement will be perceived up to about eight inches from the patient's eye; and at five or six inches from the eye an erect movement will begin. From this, the surgeon may assume that the myopia is about 7 D. [focal distance 6½ inches] and place before the eye, for the more accurate test, a concave 6 D. lens. On trying the movement of light in the pupil through this lens, it will be found at the distance of one metre to be *with* that of the light on the face. The surgeon then withdraws still farther from the patient until the direct movement becomes indistinguishable and at two metres is entirely lost. Drawing back still farther from the patient, he might in a favorable eye be able to distinguish the inverted movement in the pupil, and in this way fix the point of reversal at a distance of two metres, indicating with great accuracy an uncorrected myopia of 0.5 D.

Often, however, the distance of two metres will be found

so great that it is difficult or impossible from such a distance to be sure of the movement in the visual zone. In such a case the 6 D. lens will need to be replaced by a weaker lens as a 5.5 D., with which the erect movement will be seen to almost a metre, and the inverted movement will begin a few inches beyond that point.

If the myopia be very low, the first inspection of the pupil without a lens may show a movement of light in it *with* the light on the face. In such a case, the surgeon will draw back as far as he can readily distinguish the movement of light in the visual zone. If the movement still appears to be *with* that of the light on the face, he will place before the eye a *convex* lens, and with it determine the point of reversal as for a case of hyperopia. The final result of testing, however, will show that the myopia caused by the lens is greater than the strength of the lens, and, therefore, that some myopia must have been present before the lens was placed in front of the eye.

For example: Suppose that before reaching that distance of two metres the erect movement in the pupil becomes indistinct, and that the visual zone, where the movement must be watched, is so small that beyond this the direction of movement in it cannot be recognized with certainty. A 0.5 D. convex lens being placed before the eye is found to cause an inverted movement beyond 125 centimetres, and to confine the erect movement to within 85 or 90 centimetres of the eye. The point of reversal then, is at one metre. The amount of myopia corresponding to this is 1 D., of which 0.5 D. was the amount originally present in the eye.

Emmetropia.—On first inspection, without a lens, the surgeon sees an erect movement in the pupil, the rapidity of which indicates that if there be hyperopia it is of low degree. Drawing back from the patient's eye as far as possible, however, this erect movement still continues. He places before the eye under observation a convex lens of 1 or 2 D., and viewing the movement of light in the pupil through this lens, finds where the inverted and the erect movement come together. On measuring the distance of this point of reversal from the patient's eye, he finds that it exactly corresponds with the focal distance of the lens he has been using. That is, the lens has caused myopia just equivalent to its own strength, showing that before they passed through the lens, the rays emerging from the cornea were parallel.

Regular Astigmatism.—Whether it be known that the eye under examination is astigmatic or not, the test will proceed at first as for simple hyperopia or myopia. Sometimes if the astigmatism be high and one meridian nearly emmetropic or slightly myopic, the first inspection, without any lens, will reveal an unmistakable band of light, or that there is erect movement in one meridian and inverted movement in another, or that the movement of light in the pupil is more rapid in some one meridian than in the meridian at right angles to it, indicating that these meridians have different points of reversal, and that the surgeon is nearer the point of reversal for one than for the other, or is between the two.

But, commonly, the first appearance will give no positive indication of the presence of astigmatism, and the test goes on until a point of reversal is found. Then, on trying the movement of light and shade in different meridians, as should always be done from the neighborhood of the point of reversal, it is discovered that it is the point of reversal for only one meridian; and that for the meridian at right angles to that one, there is a distinct movement of the light either erect or inverted.

If the movement still noticeable from the point of reversal first discovered be an inverted movement—*against* the light on the face—the surgeon should bring his eye still closer to the patient until this inverted movement ceases. He will then be near the point of reversal for the meridian in which the inverted movement was before noticed, and will be able to see in the other meridian an erect movement.

Such a lens is now to be chosen and placed before the eye as will bring this point of reversal for the more myopic meridian—the point of reversal from which an erect movement is seen in the other meridian—to a convenient distance from the eye. The surgeon's eye is placed as nearly as possible at this point of reversal. Then the original source of light [which has up to this stage of the test accompanied the mirror in its movements to or from the patient's eye] is pushed away from the mirror, and while it is pushed away, the mirror is rotated and the light area in the pupil watched. This light area will now be seen to assume the band-like appearance characteristic of astigmatism.

At a certain distance this band-like appearance will be most distinct. With the source of light nearer the mirror or farther from the mirror, it will be less characteristic. The distance of the light from the mirror at which the band becomes most distinct is the distance between the two points of reversal. The surgeon's eye (with the mirror) is now at the point of reversal for the more myopic meridian, and the immediate source of light is at the point of reversal for the less myopic meridian. (See page 48.)

With the light in this position, the direction of the band is to be carefully studied and noted as the direction of one of the principal meridians of astigmatism. It is the direction of the axis of the convex cylinder that will correct the astigmatism. The other principal meridian will, of course, be perpendicular to this.

Having now fixed the direction of the principal meridians of astigmatism, the surgeon should again bring the original source of light as near to the mirror as possible, and proceed to measure the refraction, first in the one principal meridian and then in the other, just as he would measure the refraction in a case of hyperopia or myopia. The difference between the refractions of the two principal meridians being the amount of astigmatism present.

To measure the refraction in a certain meridian the light is made to move on the face, and on the retina, in the direction of this meridian by rotating the mirror about an axis perpendicular to it. Thus for the vertical meridian the light is made to move vertically by turning the mirror about a horizontal axis. For the horizontal meridian the light is made to move horizontally, by turning the mirror about a vertical axis. Great care is necessary in the higher degrees of astigmatism to make the movement conform accurately to the meridian to be tested, since any oblique movement will appear (see page 53) as though perpendicular to the band.

When the astigmatism is of very low degree, 0.5 D. or less, it becomes correspondingly difficult to distinguish between the points of reversal for its principal meridians. The band-like appearance of the light in the pupil becomes less characteristic, and there is no space between the two points of reversal where an erect movement can be obtained in the direction of one meridian, and a reverse movement in the direction of the meridian perpendicular to it. In these cases, the astigmatism is to be recognized by the fact that when, near one point of reversal, the movement in one meridian has become indistinguishable, it can still be perceived in the other principal meridian. And, if the surgeon places his eye at the point of reversal for the more myopic meridian and pushes the source of light a little away from the mirror, the erect movement in the meridian of less myopia, and absence of movement in the more myopic meridian become most distinct. It is upon this behavior of light in the pupil under these conditions that the diagnosis of the very low degrees of astigmatism must principally rest.

The final test in any case will be made with the points of reversal brought together, usually at a distance of 1 metre or less. To do this, it will be necessary to place before the eye such a cylindrical lens as will correct the astigmatism, together with the spherical lens which will bring the point of reversal to the desired distance. With these lenses before the eye, the test is again applied. If the light in the pupil is found to move *with* the light on the face, the surgeon withdraws to a greater distance until that movement becomes indistinct. If the movement in the pupil is found to be *against* that of the light on the face, the surgeon approaches the patient until the movement becomes indistinct. The apparent movement is to be carefully inspected from the point of reversal and from a little within and a little beyond it.

If it is found that the reversal occurs at the same distance from the eye for all meridians, the cylinder chosen is known to be correct, both as to strength and as to the placing of its axis; and the distance of this point of reversal from the eye indicates the amount of myopia which the spherical lens employed has caused, or has left uncorrected.

If, however, the movement of light is found to cease in some meridian, but to continue (either direct or inverted) in a meridian at right angles thereto, it becomes evident that the cylinder chosen does not perfectly correct the astigmatism. If the astigmatism, thus found to remain, has the same principal meridians as those already fixed upon, the direction of the axis of the lens is correct, but its strength is not exactly right. Whether the strength needs to be increased or to be diminished will appear from the fact that the more myopic meridian continues to be the more myopic; or that what was originally the less myopic meridian has become the more myopic.

If the astigmatism remaining after the cylindrical lens has been placed before the eye has principal meridians that do not correspond with those for which the lens is placed, the placing of the lens is incorrect, and the direction of its axis needs to be slightly varied, until the remaining astigmatism disappears or its direction corresponds with that of the lens before the eye.

Where the cylindrical lens before the eye is of the right strength or is too weak, its axis needs to be turned slightly toward the axis of a similar cylinder which would correct the remaining astigmatism. If the cylindrical lens already before the eye is too strong, its axis needs to be turned toward the axis of a cylindrical lens of the opposite kind that would correct the astigmatism.

The effect of such combinations of cylindrical lenses may be more fully understood by the study of the writer's paper upon "The Equivalence of Cylindrical and Spherocylindrical Lenses" in the *Transactions of the American Ophthalmological Society* for 1886, page 268, or "Some Remarks on the Refractive Value of two Cylinders" by Carl Weiland *Archives of Ophthalmology*, 1893, p. 435, and 1894, page 28.

When the meridians of any remaining astigmatism have thus been made to conform to the direction of the cylindrical lens before the eye, this remaining astigmatism has to be corrected by a change in the strength of the cylindrical lens.

For example: Suppose an eye having a compound hyperopic astigmatism corrected by $+$ 1. sph. $\bigcirc$ $+$ 1. cyl. axis 95°. The first inspection of the movement of light in the pupil shows a movement *with* that of the light on the face in all meridians, and a difference in the rate of movement in the different meridians so slight as probably to escape notice. A convex 3 D. spherical lens will cause the movement in the pupil to be *against* the light on the face in all meridians when the eye is viewed from a greater distance than one metre. But it will also be noticed that the light moves more swiftly from side to side than it does upward and downward.

If now the surgeon brings his eye closer to the patient, when the distance of one metre is reached, the movement of the light from side to side becomes indistinguishable, while there is still a very distinct movement *against* the light on the face upward and downward. Approaching still closer, the movement from side to side is seen to be *with* the movement of the light on the face, the inverted movement still continuing in the vertical meridian. The movement horizontally *with* the light on the face, at first very rapid, grows slower as the patient's eye is approached, and the movement—*against* the light on the face—in the vertical meridian grows more rapid, until at a distance of one-half of a metre, the movement in the vertical meridian becomes indistinguishable, while there is a very clear movement of light, *with* the light on the face, from side to side.

The point of reversal for the more myopic meridian (more *myopic* with the lens) has now been reached, and the surgeon keeping his eye at this position pushes the source of light away from the mirror. As he does so, the area of light in the pupil assumes more and more the appearance of a distinct vertical band, readily moved from side to side, but without apparent movement in the direction of its length. This band continues to become more distinct, until the original source of light is one-half metre from the mirror, and the immediate source consequently one-half metre back of the mirror, and one metre from the patient's eye —at the point of reversal for the less myopic meridian. In this position careful observation will show that the band of light in the pupil is not exactly vertical, but has the direction corresponding to the more myopic meridian of 95°. The principal meridians then are located at 5° and at 95°.

Having determined this, the light is brought back as close to the mirror as possible, and the point of reversal for the 95° meridian is determined. To do this it may be advisable to change the convex spherical lens before the eye, but whatever lens is employed, from the results obtained with it, the surgeon deduces the fact that in that meridian the refraction of the eye is hyperopic 1. D. He then proceeds to measure in the same manner the refraction of the eye in the other principal meridian, finding with the convex 3 D. lens that this point of reversal is at one metre, and its refraction, therefore. hyperopic 2. D. The difference between these meridians will be 1. D., the amount of astigmatism present.

To make the final determination, there should be placed before the patient's eye the 1. D. convex cylinder with its axis at 95° and a 2. D convex spherical lens; with which the point of reversal for all meridians will be found to lie one metre from the eye. If, in the placing of the cylinder, its axis is made to not correspond exactly with the meridian of least hyperopia, there will be found by this test a remaining astigmatism of low degree. Suppose, through carelessness or inaccuracy in the earlier observation, the axis of the cylinder should be placed at 105° instead of at 95°, the remaining astigmatism then would be found to be such as would be corrected by a convex cylinder with its axis at about 70°. But on turning the cylinder before the eye 10° in that direction, that is, to its proper direction at 95°, this remaining astigmatism would disappear. If, however, instead of the 1 D. cylindrical lens, a lens of 1.5 D. had been placed with its axis at 105°, there would remain an astigmatism which might be corrected by a concave cylinder with its axis at about 70°, and the turning of the cylinder before the eye 10° in that direction [to 95°] would cause the remaining astigmatism to so change that its meridians would be at 5° and 95°, where a measurement of it would reveal the fact that the cylindrical lens employed was 0.50 D. too strong.

Aberration and Irregular Astigmatism.—The difference in the refraction of different parts of the pupil is to be ascertained by measuring the refraction for each part separately, just as though it were a case of simple hyperopia or myopia, care being taken to confine each observation strictly to the little portion of the pupil the refraction of which it is desired to ascertain.

The amount of aberration or irregular astigmatism that is thus ascertained is of some scientific interest, and occasionally of practical importance as bearing on the prognosis of conical cornea, or of the changes of refraction in the lens which precede cataract.

Generally, however, the important practical point about aberration or irregular astigmatism is its distribution. For practical purposes, the surgeon desires to ascertain which part of the pupil is free from any such defect, as that part will furnish the best visual zone; and by what lenses that visual zone can be made most useful to the patient. The need for careful study to develop these points is sometimes great. Figures 22 and 23 represent the appearances brought out by thorough investigation of a case

FIG. 22.   FIG. 23.

of considerable astigmatism, coincident with equally pronounced positive aberration. Without careful study of the visual zone at the proper distances, it would have been easy to set the case down as one of aberration, and to have overlooked the regular astigmatism entirely. If the aberration encroaches decidedly upon the area of the pupil as determined by a moderate light, it may be necessary to give a correcting lens, for use at near work and on exposure to bright light, different from the one required when the pupil will be somewhat larger. Or the surgeon may need to caution the patient that under certain conditions of light he must expect the correcting glasses to give slightly imperfect vision. (See also page 68.)

A fact to be borne constantly in mind in the application of skiascopy is, that it is not the high degrees of aberration and irregular astigmatism that are of most practical importance, requiring the surgeon to take into account their bad effects. More frequently it is the slight imperfections of this kind situated within the portion of the pupil used for accurate vision that need to be recognized and taken into account, when prescribing glasses or in giving an opinion as to the value of glasses. These low degrees of imperfection are to be recognized and studied only when the surgeon's eye is close to the point of reversal for the visual zone, after the effects of hyperopia, myopia and regular astigmatism have been excluded by placing the proper glasses before the patient's eye.

The investigation of aberration and irregular astigmatism is the last step in skiascopy. A step very frequently not taken, yet essential to complete certainty and accuracy in the objective measurement of refraction. In a small proportion of cases, it will lead to modification of the glasses previously selected as best, and in a much larger proportion of cases it will discriminate sharply between the lenses which really best correct the ametropia and others which appear to give equally good, or almost equally good, subjective results.

To carry it to completion will often require more time and effort than has been necessary for all other parts of the skiascopic examination. Nor is this strange, for it often includes the measurement of hyperopia or myopia, perhaps with astigmatism, in two or more areas of distinctly, though slightly, different refraction. It is not a distinct application of the test, but its application to separate parts of the pupil, instead of to the pupil as a whole. It requires no special directions and cannot be much elucidated by examples. It is to be mastered by a full understanding of the optical principles of the test, Chapter II, strict observance of the conditions of accuracy set forth in Chapter III, and the exercise of the needful care and patience.

Measurement of Accommodation.—The objective determination of the nearest point for which the eye can be focused is possible only by skiascopy. It is sometimes of importance as in cases of suspected cycloplegia in children, or others for whom the subjective test cannot be relied on. In determining the condition of the accommodation in an eye with imperfect vision, or in recognizing any slight remaining accommodation after the use of a mydriatic, the test is also of practical value.

To make it, the surgeon first ascertains the refraction of the eye, and then places before it such a lens or lenses as will correct astigmatism and bring the point of reversal to a distance of one metre or a little less. He then places himself at this distance from the patient; and directs the patient to fix his gaze upon some object on the farther side of the room, in such a position that the visual axis of the eye under examination shall pass as close as possible to the surgeon's eye. The point of a finger or pencil is then held close to the patient's eye, about the near limit of convergence and in the visual axis, so that the direction of the visual axis shall not be materially changed during the test.

The patient is then directed to look first at the object across the room, then at the point of the pencil close to his eye. The surgeon by watching his other eye can ascertain whether the movements of convergence are really executed. Very strong convergence being impossible without strong accommodative effort, if any power of accommodation remains, the eye will be seen to grow more myopic when the pencil is looked at, and less myopic when the distant object is fixed, an inverted movement of the light in the pupil becoming apparent on "fixing" the near object and disappearing on "fixing" the other.

To measure the amount of accommodation the surgeon may approach the observed eye until the point of reversal is reached for the eye during very strong convergence; or the lens before the eye may be modified in the direction of weaker convex or stronger concave until the point of reversal is brought, with a new lens and the accommodation, to the same distance as it was brought by the original lens without accommodation. The difference between the lenses in this case represents the amount of accommodation present.

For example: in a case of hyperopia 2 D. to ascertain if accommodation were present a convex 3 D. lens would be placed before the eye. This would bring to one metre the point of reversal of the eye with its accommodation relaxed. The surgeon at a little less than a metre could get through this glass an erect movement of the light in the visual zone when the patient was looking across the room. If now, on looking at the point of a finger held two inches in front of the eye, the movement becomes distinctly inverted, the light in the pupil moving against the light on the face, it is known that accommodation is present. In place of the 3 D. lens a weaker convex may be substituted, and if the strong convergence of the visual axes still brings an inverted movement of light in the pupil, a still weaker lens for this. In this way if it be found that with the 1 D. lens the patient is able by strong accommodative effort to bring the point of reversal to just one metre, the difference between the 3 D. lens and the 1. D.=2. D. will be the amount of accommodation present.

Instead of changing the lens, the surgeon can approximately estimate the amount of accommodation by bringing his eye closer to that of the patient and finding the new position of the point of reversal caused by the exertion of the accommodation. Where much accommodation is present, such an approximate determination should first be made, but it is liable to the inaccuracies attendant on any skiascopic determination at a short distance.

CHAPTER VII.

PRACTICAL APPLICATION WITH THE CONCAVE MIRROR.

The Source of Light.—The room should be thoroughly darkened; and to secure this, it is well to have the original source of light shaded. This light will, however, usually be back of the patient, and, except for the determination of the principal meridians of astigmatism, the farther it is behind the patient, the better. Hence, the shading of the light is not essential, as it is with the plane mirror. It is also of less importance that the original source of light should be small. Still the separation of light and shade should be as sharp as possible, so that the opaque shade with an opening opposite to the brightest part of the flame will be found serviceable. The opening in the shade may be two or three centimetres in diameter, so long as the original source of light is a metre or more away from the mirror. But when this is brought near the mirror to bring out the band of light in astigmatism the opening should be one centimetre in diameter or less.

The surgeon places himself with his eye one metre from that of the patient. On throwing the light upon the patient's face with the mirror, it is found that the area of light on the face moves with the mirror just as in the case of the plane mirror. The same reflections from the cornea and trial lenses are to be recognized and guarded against; and within the pupil lies a similar portion of red fundus reflex, bounded by shadow, which is the subject of observation during the test. If, however, the light in the pupil be seen to move *with* the light on the face, the eye is myopic more than 1. D. If the light in the pupil be seen to move *against* the light on the face, the eye is hyperopic, emmetropic or less than 1. D. myopic. (See pages 23-26.)

Hyperopia.—If the mirror be rotated about a vertical axis from right to left, the area of the light in the pupil will be seen to move from left to right, that is, *against* the mirror and *against* the light on the face. This is really an erect movement, we know from the demonstrations as to the real direction of the movement of the light on the retina in Chapter II. The difference between *erect* movement and *movement with* the light on the face must be borne in mind. With the concave mirror, the one is just the opposite of the other. The movement *with* the light on the face being an *inverted* movement; and the movement in the pupil *against* the light on the face the *erect* movement.

As with the plane mirror, the movement will be swift if the hyperopia be of low degree; slower, if of higher degree. The convex lens is now to be placed before the eye, the swiftness of the movement in the pupil being the guide to the strength probably required. If the observer is not able to judge by this movement, let him at first employ in succession lenses that differ considerably in strength, as the 2, 4, and 6. D., increasing the strength as long as the movement in the pupil is *against* the movement on the face.

When a lens is reached that causes movement of light in the pupil *with* the light on the face, slightly weaker lenses are to be tried until the two consecutive lenses are found, of which one gives the movement *against* the light on the face and the next stronger causes movement *with* the light on the face. Between these two lies the lens strength which would bring the point of reversal to the surgeon's eye. The surgeon's eye being one metre from the patient, this is the lens which would cause 1. D. of myopia; and by subtracting 1. D. from its strength the hyperopia of the eye is obtained.

For example : Suppose hyperopia of 4. D. The light in the pupil will move *against* the light on the face at the first inspection, and also with the convex 2. D. and 4. D. lenses. With the convex 6. D. it is found to move *with* the light on the face. On trying the convex 5. D., the movement is indeterminable. With the 4.75 D., it is very rapid, but still *against* the light on the face. With the 5.25. D., it is equally rapid, but *with* the light on the face. The lens strength between the two, or the 5. D., is then the one which causes 1. D of myopia ; and 5. D., the strength of lens, minus 1. D., the myopia caused by it, leaves 4. D., the lens strength required to correct the hyperopia present.

Myopia.—In the mass of cases the inspection without a lens will show the movement of light in the pupil *with* the movement of light on the face, indicating the point of reversal is between the surgeon and the patient. When this is the case, concave lenses are to be tried, their strength being indicated by the rate of movement; or, if this be not a sufficient guide, they may be tried in series with an interval of about 2. D., until one is found which causes the light in the pupil to move *against* the light on the face.

As the point of reversal is thus brought farther from the eye of the patient and nearer to the observer's eye, the light area in the pupil becomes more brilliant, and its movement more rapid. When the lens has been found which causes the light in the pupil to move *against* the light on the face, slightly weaker lenses are to be tried until it has been certainly ascertained which is the weakest lens that will cause the movement *against* that of the light on the face, and which is the strongest lens that still allows movement in the pupil *with* the light on the face. Between these two lies the lens-strength which leaves the eye 1. D. myopic. This lens-strength added to 1. D. will give the total myopia present.

For example : Suppose the myopia present to be 8.5 D.

The movement in the pupil without any lens will be very slow, and the light areas round and dim. Judging from this appearance the first lens tried may be the concave 5. D. With it, the light in the pupil will appear more brilliant and its movement will be more rapid, but it will still be *with* the movement of the light on the face. Next, the concave 8. D. will be tried. The movement of light will be found still more rapid, but now *against* that of the light on the face. With the concave 7. D., it will be found equally rapid, but *with* the light on the face. With the 7.5. D. it will not be distinguishable. Hence, the 7.5. D. lens leaves 1. D. of myopia still uncorrected, and this added to the 7.5 D. corrected by the lens gives 8.5. D., the total myopia present.

If the myopia be of low degree, the test without a lens will show either no distinguishable movement of light in the pupil [for myopia of 1. D.], or movement in the pupil *against* the movement of light on the face [for myopia of less than 1 D.]. In the former case the test is to be repeated with very weak convex and concave lenses [0.25. D. or 0.50. D.]. The convex will give a movement of the light in the pupil *with* the light on the face, and the concave movement *against* the light on the face.

If the movement is found to be *against* the light on the face to start with, the convex lenses are to be tried, commencing with a 1. D. lens, which will cause the movement *with* the light on the face, and will show, therefore, that the refraction is myopia and not emmetropia, or low hyperopia. The weaker lenses are then to be tried, and the one which causes 1. D. of myopia thus ascertained. Since this lens is added to the myopia of the eye to cause 1. D. of myopia, it must be subtracted from 1. D. to find the amount of myopia originally in the eye; the difference between it and 1. D. being the myopia present.

Thus, in a case of myopia of 0.50. D. the light will be

found to move *against* the light on the face, without any lens, or with a 0.25. D. convex. But will be found to move *with* with the light on the face, with a convex 1. D., or 0.75. D.; and with an 0.50. D., the movement should be indistinguishable. The convex 0.50. D. then causes 1. D. of myopia, and subtracting it from 1. D. leaves 0.50. D., the degree of myopia previously existing in the eye.

Emmetropia.—In emmetropia, on the first trial, the light in the pupil is found to move rapidly *against* the light on the face. With convex lenses it is found that the 0.75 D., or anything weaker, still allows this movement *against* the mirror. But the 1.25 D., or anything stronger, causes motion in the pupil *with* the light on the face; and that the convex 1. D. causes no perceptible movement. Hence, the convex 1. D. lens causing 1. D. of myopia, the eye without a lens must be emmetropic.

Regular Astigmatism.—The test beginning as for simple hyperopia or myopia, as the point of reversal for one of the principal meridians is brought near the observer's eye the movement of light becomes notably more rapid in one meridian than in the other, indicating the presence of this form of ametropia. When this is recognized, the lenses used are to be such as give a movement of light in the pupil *with* the light on the face in all meridians. Thus, if the eye has been hyperopic, the convex lenses used before the eye must be increased in strength until the movement *with* the light on the face occurs in all directions. Or, if the eye is myopic, the increase of strength in the concave lenses must stop so soon as any movement is seen in the pupil *against* the movement of light on the face. And the lens which causes this must be replaced by a weaker one that just allows movement with the light on the face in all meridians.

The lens aimed at is the one which will bring the point of reversal for the least myopic meridian just to the surgeon's eye, one metre from the patient. If this is exactly attained, there will be in that meridian no perceptible movement of light and shadow, but the movement in the other principal meridian will still be *with* that of the light on the face.

When this lens has been found, the original source of light, which, up to this time has been kept at as great a distance as possible from the mirror, is to be brought closer to the mirror, so that the image of it formed at the conjugate focus in front of the mirror—the immediate source of light—will be removed farther from the mirror and closer to the patient's eye.

The lens before the patient's eye brings the point of reversal for the least myopic meridian to the eye of the surgeon, and necessarily places the point of reversal for the more myopic meridian somewhere between the surgeon and patient. The object of bringing the original source of light nearer to the mirror is to carry the immediate source of light to the point of reversal for this more myopic meridian. As the light approaches its proper position, the area of light in the pupil becomes more and more band-like, being most distinctly so when the immediate source of light corresponds with the point of reversal for the more myopic meridian.

When this is attained, the direction of the band is to be carefully noted as indicating the direction of the principal meridian of least myopia. This direction having been determined and recorded, the original source of light is again moved as far away from the mirror as possible, and measurement of the refraction in the least myopic meridian completed as for a case of simple myopia or hyperopia.

Then, the lenses are so changed as to bring the point of reversal for the more myopic meridian to the surgeon's eye, one metre distant from the patient; and the lens that is found to do this, shows by the addition of 1. D. if concave, or the subtraction of 1. D. from the strength if convex, the amount of myopia or hyperopia in the second meridian. The difference between the two meridians is the amount of astigmatism.

When this amount of astigmatism has thus been ascertained, the cylindrical lens correcting it is to be placed before the eye and with it, the spherical lens, which will bring the point of reversal to a distance of one metre. The trial is then repeated, and if the point of reversal be found at the surgeon's eye for all meridians of the pupil, the determination already made is accurate. If, however, there be found distinct movement in the visual zone in some one direction while movement in the principal meridian perpendicular thereto is abolished, the cylinder selected does not perfectly correct the astigmatism.

If this movement be in one of the principal meridians as previously determined [in the direction of the axis of the cylinder placed before the eye, or at right angles to that axis] the cylinder has been placed in the proper direction, but is too strong or too weak; and its strength must be diminished or increased, according to the indications of the movement. If, however, the movement appears to be in a meridian different from either of the principal meridians as at first determined [different from the direction of the axis of the cylindrical lens before the eye, or the principal meridian at right angles to that axis] the axis has not been properly placed—does not conform exactly to the direction of the principal meridian.

If this is the case and the cylindrical lens before the eye is of the right strength or too weak, its axis needs to be turned slightly toward the axis of a similar cylinder, which will correct the remaining astigmatism. If the cylindrical lens already before the eye is too strong, its axis needs to be turned toward the proper position for the axis of a cylindrical lens of the opposite kind that would correct the astigmatism. Such a change in the direction of the axis of the cylinder is to be made, and the test repeated until the correction of any remaining astigmatism conforms exactly with the direction of the lens before the eye. This remaining astigmatism must be corrected by a change in the strength of the lenses employed.

For example: Suppose an eye to have compound hyperopic astigmatism corrected by $+$ 4. sph. $\subset$ $+$ 2. cyl. axis 90°. The first inspection of the pupil shows the light moving *against* the light on the face in all meridians. Convex lenses 2. D. and 4. D. placed before the eye show the same thing. Convex 6. D. shows the light moving *against* the light on the face from side to side, but *with* it in a vertical direction. It thus becomes evident that astigmatism is present. Still stronger convex lenses are to be tried. The 8. D. lens shows movement in the pupil *with* the light on the face in all meridians. The 7. D. lens shows movement very indefinite or indistinguishable in the horizontal meridian, but clearly *with* the light on the face in the vertical meridian. This lens then brings the point of reversal for the less myopic [more hyperopic without the lens] meridian to the surgeon's eye.

The next step is to bring the original source of light closer to the mirror, so as to cause the immediate source of light to fall at the point of reversal for the more myopic [less hyperopic] meridian, which will now be one-third of a metre from the patient's eye. To do this [supposing that the mirror has a focal distance of one-quarter of a metre, ten inches] it will be necessary to bring the source of light to within two-fifths of a metre of the mirror. That is, the immediate source of light to be at one-third of a metre from the patient, must be at two-thirds of a metre from the mirror corresponding with 1.5 D. of the focusing power. The total focusing power of the mirror being equal to 4. D., the light must be so placed that the divergence of its rays will correspond to 4–1.5=2.5 D. That is, the light must be two-fifths of a metre from the mirror. When the light is in this position, the area of light in the pupil will assume the most distinct band-like appearance running in the direction of the principal meridian of least myopia [greatest hyperopia], in this case horizontal.

Having thus determined the direction of the principal meridians, one being known from the direction of the other, the original source of light is again placed back of the patient as far as possible, and the refraction in the horizontal meridian carefully tested by trying first the 6.5 D. spherical lens, and then the 7.5 D. spherical lens before the eye; the former of which shows the movement in a horizontal meridian *against* the light on the face, and the latter a movement in the same meridian *with* the light on the face, thus fixing the refraction of that meridian as 7. D.–1. D.=6. D. of hyperopia.

Weaker convex lenses are then to be tried, until it is found that with the 5.5 D. the light moves *with* the light on the face in the vertical meridian, and with the 4.5 D. it moves against the light on the face in the vertical meridian, while the 5. D. gives no distinguishable movement in that meridian; showing that 5. D.–1. D.=4. D. is the amount of hyperopia in the less hyperopic meridian. The difference between the two then is found to be 2. D., the amount of regular astigmatism present.

The surgeon will then place before the patient convex 5. D. spherical with convex 2. D., cylindrical, axis vertical. And on again trying the test, will find that he is at the point of reversal for all meridians. But, if on placing the cylindrical lens he make a slight error in the direction of its axis, placing it say at 5° one side from the vertical, he will find on testing the eye some appearance of astigmatism with its axis inclined several degrees in the other direction from the vertical. And, to get rid of this astigmatism, he has to move the axis of the cylindrical lens to its proper position, pushing it toward the axis of a convex cylinder that would be required to correct this remaining astigmatism.

If the case be one of slightly myopic, or high mixed astigmatism, the first inspection may show a movement *with* the light on the face in one direction, while the movement is *against* the light on the face in the other meridian. This, of course, will indicate at once the presence of astigmatism. The fact, that it may occur, makes it important that the first observation on the pupillary movements should include the movements in different meridians.

With the concave mirror [the immediate source of light necessarily lying as far in front of the mirror as its principal focus, or even farther] if the astigmatism be of quite low degree, when the least myopic point of reversal is at the surgeon's eye, the more myopic point of reversal will be at the *immediate* source of light, or even closer to the mirror without any change in the position of the *original* source. Thus, the most distinct band-like appearance of the light in the pupil, the clearest difference between the movement against the light on the face in one meridian and the indefinite movement in the other merdian will be attained without bringing the original source of light any nearer to the mirror than its usual position. This must be born in mind for low degrees of astigmatism.

Aberration and Irregular Astigmatism.—With the concave mirror, and the need of bringing the point of reversal to a fixed distance from the patient's eye, the measurement of the amount of aberration and irregular astigmatism becomes very much more tedious and difficult than with the plane mirror, though not impossible. It is, however, not difficult to detect the presence of such defects; and to ascertain which portion of the pupil they occupy, and which portions being comparatively free from them are available as a visual zone. As to the importance of such a study of the pupil, what has been said in the chapter on the plane mirror (page 83) will equally apply here.

Measurement of Accommodation.—With the aid of lenses, usually concaves, the near point of accommodation may be brought to the required distance of one metre from the eye, and the amount of accommodation thus measured. The arrangement of the patient's and surgeon's eyes, and of the points to be looked at, is the same as that described in connection to the measurement of accommodation with the plane mirror. It is, of course, impossible to make the approximate determination of the accommodation with the concave mirror by the surgeon approaching the eye of the patient. He must rely entirely on a change of lenses to bring the point of reversal to the fixed distance of one metre.

CHAPTER VIII.

GENERAL CONSIDERATIONS.

Apparatus.—In the chapter of the Conditions of Accuracy, something has already been said as to the apparatus by which these conditions are best complied with. The two requirements to meet which the apparatus for skiascopy is to be adapted, are that it shall furnish the conditions necessary to the greatest accuracy, and that it shall facilitate the finding of the lens that will bring the point of reversal to the surgeon's eye.

The Mirror.—As has been stated, the essential point in the mirror is the sight-hole small enough and free from reflections. This may be obtained by having the glass thin, if the sight hole is cut through it, having its margin free from chipping, beveled as little as possible, and thoroughly blackened with a dead black.

If the sight hole is not cut through the glass, but is merely an aperture in the silvering, the glass may be much thicker and there is no ground glass to deal with. The difficulty with such a mirror is in keeping the exposed glass at the sight hole clean. Unless great care is taken in preserving it from dust, and carefully removing any that falls upon it, there will be a ring of dust in the periphery of the sight hole, which will irregularly reflect more light than would the ground glass edge of the perforated sight hole. And, it is difficult to keep this space entirely clean without chipping into the back of the mirror in such a way as to cause annoying reflections. But, however difficult, it is important to have the sight hole free from reflections.

THE MIRROR.

The size of the mirror will depend somewhat upon the purpose for which skiascopy is to be used. If the mirror is to be employed to measure refraction of all kinds, to show the movement of light in the pupil with high uncorrected hyperopia or myopia, it must be large, to give the range of movement for the immediate source of light that is necessary to render evident the direction of movement in the pupil, when that movement is slow and the illumination of the area is comparatively feeble.

The disadvantage of a large mirror is that it gives a large area of light on the face, especially when as with the plane mirror, the original source of light is brought close to it. And in this large area of light on the face only the light reflected by a small portion of the mirror immediately surrounding the sight hole is of any use when the point of reversal is near to the surgeon's eye [see page 29 for discussion of limits of the part of the retina visible in the pupil]. With a small mirror, making a small area of light on the face, it is easier to keep this upon the eye than it is to keep the similar limited portion of a large area properly directed.

On this account, where skiascopy is used, after an approximate estimate of the refraction has been made by the ophthalmoscope or other means, quite a small mirror is found convenient. By a large mirror is meant one from 35 to 50 mm. in diameter. By a small mirror is meant one under 20 mm. in diameter. The mirror, or, at least, the opaque back that carries it, cannot be well reduced to less than 20 or 25 mm., because, if smaller than this, it will admit light to the eye from the original source, through the space around the mirror; and such light, though not so annoying as a reflection at the sight hole, is a serious hinderance in the application of the test. The mirror plate then must be large enough to shade the eye.

A large mirror having a metal cap with an aperture of

from 10 to 15 mm. in diameter, that can be slipped before the face of the mirror, or turned back at pleasure, will answer for all sorts of testing. Such a mirror[1] is shown in figure 24. As already indicated in Chapter III, the sight hole should be about 2 mm. in diameter.

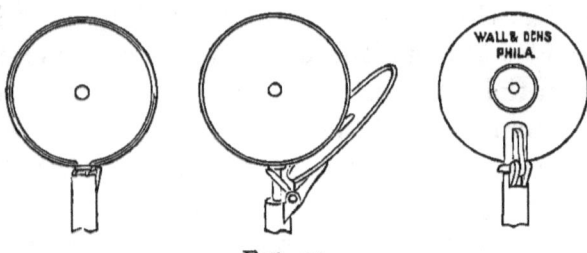

FIG. 24.

The handle of the mirror should be rather thick, so that a very slow even rotation can be secured; for, as the point of reversal is approached, the magnified movement in the pupil becomes so rapid that only by moving the mirror more slowly, and making excursions of very slight extent, can this apparent motion in the pupil be readily followed. This difficulty of causing the immediate source of light to move slowly enough is diminished in proportion as the immediate source of light is brought closer to the mirror.

The Shade.—The shade that covers the original source of light should extend far enough above and below the flame to prevent the escape of any considerable amount of light into the room. Where an argand burner is used as the source, a cylindrical shade should be 20 to 25 cm. long, with a diameter 6. or 6.5 cm., slightly greater than that of the chimney used, so as to allow a free current of air between the shade and chimney and thus diminish the heat from the flame. An asbestos shade has been proposed by Dr. J. Thorington (*Ann. of Ophthalmology and Otology*, 1895, p. 5) on account of intercepting better the heat of the flame.

[1] Made at my suggestion by Wall & Ochs, of Philadelphia. Another form is described by Dr. James Thorington, *Philadelphia Polyclinic*, 1893, page 320.

The aperture of about 5 mm. for the plane mirror, or larger, for the concave mirror, should be opposite the brightest part of the flame, which ought to be broad enough to allow of slight change of position of the surgeon with reference to it, without its becoming hidden by the shade.

The Lenses.—Ordinarily these are taken from the trial case and placed in a trial frame before the eye. It is important to have them clean and comparatively undamaged by scratching. The trial frame should be such as to support the lenses well up before the eye and with their centres before the centres of the pupils. They must also be far enough away from the face to escape the touching of the lashes, and to prevent the condensation of moisture upon them. The interruption of the red reflex from the pupil by such an occurrence prevents the satisfactory application of the test, and may be quite puzzling, because the reason for the obscuration is not immediately apparent; and it may be ascribed to opacities within the eye.

Support of Lenses.—The trial frames have the advantage over other supports for lenses to be presently mentioned, that they keep a constant position with reference to the patient, so that a slight movement of the patient's head does not carry his eye away from the centre of the lens to its periphery or beyond.

When the surgeon has learned to estimate by the rapidity of movement of the light in the pupil, the amount of ametropia remaining uncorrected, by following the plan here laid down of considerable intervals between the lenses until an approximation of the required lens has been made, the number of changes of lens for any case is not necessarily great. So that for any one who does not employ skiascopy on large numbers of patients daily, the trial frame and lenses will be found entirely satisfactory.

Special series of lenses mounted in revolving disks have been arranged by Haines (*Ophthalmic Review*, 1886,

p. 282), Burnett, Doyne, Couper, (*Trans. Am. Ophthalmol. Soc.*, 1888, p. 223), Würdemann, and others, to save time by facilitating the changes to the lens required. Some of these have been designed for the patient to make the change of lens under the direction of the surgeon, and others to give the surgeon himself control of their movements.

One of the simplest arrangements is that described by Würdemann (*American Journal of Ophthalmology*, 1891, page 223), shown in figure 25. The lenses are inserted in a sheet of hard rubber which the patient holds by the handle, bringing before his eye the lens the surgeon may indicate.

In an instrument suggested by the writer the disk is rotated by a rod one metre long and attached by a universal joint, so that it drops out of the way when not in use.

The lens series runs from 7 concave to 7 convex spherical, with 0.5 D. intervals, requiring to be supplemented by lenses in the trial frame, for high hyperopia and myopia, or astigmatism.

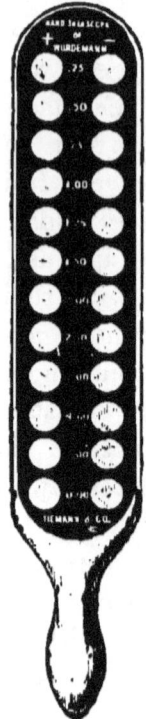

FIG. 25.

An ingenious piece of apparatus having a complete series of lenses, both spherical and cylindrical, arranged for the purpose, is described by Lambert (*Trans. Amer. Ophthalmol. Soc.*, 1894, p. 196). It has the lenses arranged in two disks for the spherical lenses, and detachable slides for the cylinders, enabling the surgeon to reach the lenses wanted quickly. To be compelled to run over the lens series to find the one sought, would be a way of consuming time rather than saving it. Other forms of elaborate apparatus for the purpose have been suggested by Sureau of Paris, and one by Perkins and Tait of Philadelphia. A series sufficient for the approximate testing of the majority of eyes may save time where many are to be tested, especially if the concave mirror be employed. The writer, using habitually the plane mirror, has discarded all special forms of apparatus, and depends on the trial frame and test lenses.

Meridian Indicators.—In working with lenses in the graduated trial frame one may refer to its graduation to ascertain the direction of the bands of astigmatism. But in the darkened room this is not convenient. To meet this want, Thorington (*Medical News*, March 3, 1894) and Prince (*Ophthalmic Review*, July, 1894) have suggested disks specially graduated for the purpose. The former, figure 26, is called an *axonometer;* the latter, figure 27, an *inclinometer.*

FIG. 26.

FIG. 27.

A Distance Measure.—Where the concave mirror is employed, the distance remaining fixed throughout the test, it is only necessary that the surgeon should properly place himself at the beginning, and retain his position. He can then dismiss the consideration of the distance, or provide for it by the addition of 1. D. to the concave spherical lens or the subtraction of 1. D. from the convex spherical lens that brings the point of reversal to his eye.

With the plane mirror no measure is necessary where the test is used only to approximate the refraction, the surgeon soon learning to guess at the distance closely enough to be within 0.25. D. of the amount of myopia present with the lens fixed upon. But for exact measurement it is convenient to have something to measure from the patient's eye to the surgeon's. This may be either a tape attached to the trial frame or lens disk (Burnett), and picked up and held to the surgeon's eye when the test is completed, or the ordinary metre stick. In either case, it is convenient to have the measure graduated in dioptric focal lengths, as described by the writer in the *Medical News*, June 27, 1885. The graduation should begin from the end nearest to the patient's eye.

Mydriatics.—In making any measurement that is to have definite significance, the first essential is that the quantity to be measured should be fixed. When the refraction of the eye varies from moment to moment, it is impossible to make a valuable measurement of it by any method. When it is liable to vary from moment to moment there is a liability to error, due to such variations. Believing that when consulted as to an error of refraction or its effects, the ophthalmic surgeon should ascertain its degree with exactness and certainty, the writer is accustomed to employ a mydriatic, so as to secure complete paralysis of accommodation in the great majority of cases under fifty years of age.

While any of the true mydriatics, atropin, daturin, duboisin, hyoscyamin, or scopolamin, will give a satisfactory paralysis of accommodation, if a mydriatic is used solely for diagnostic purposes, homatropin should be selected, on account of its briefer period of recovery. Properly used it is, for all practical purposes of diagnosis, in the great mass of cases as reliable as any mydriatic we possess.

To secure paralysis of accommodation, therefore, four to six drops of a three per cent. solution of homatropin hydrobromate are to be instilled in the eye at intervals of five minutes, about an hour before the test is to be applied.

To apply skiascopy with the greatest ease requires a pupil moderately dilated. Like other methods for the measurement of refraction, it will not give as accurate results if the pupil be narrow; and, on account of the aberration and irregular astigmatism that usually exist near the margin of the lens and cornea, the wide dilatation of the pupil introduces factors of confusion. The need then for a dilated pupil is about the same with skiascopy as for the use of the refraction ophthalmoscope or the test lenses, except that skiascopy is slightly more at a disadvantage when the pupil in the dark room is less than four millimetres in diameter. When this is the case sufficient dilatation can be obtained, with the least inconvenience to the patient, by placing in the eye a drop of a two or four per cent. solution of cocaine, thirty to fifty minutes before using the test.

Relative Advantage of Plane and Concave Mirrors.— The difference in methods of using most efficiently the plane and concave mirrors have caused most surgeon's to habitually employ the one or the other, and to depend upon it almost entirely for most practical purposes. Either properly used will meet the requirements of practice.

In astigmatism the plane mirror is capable of determining with greatest accuracy the meridian of greatest myopia, but not the meridian of least myopia. On the other hand, the concave mirror fixes with greatest accuracy the meridian of least myopia, but not that of greatest myopia. But for regular astigmatism—the astigmatism that can be corrected by a cylindrical lens, or by any combination of cylindrical lenses—the principal meridians are always perpendicular to each other. So that for practical purposes, it is only necessary to accurately locate one of them; and it is a matter of indifference which one this shall be.

In *positive aberration*, the focusing of the light upon the retina being such that the light area has the sharpest outline when the immediate source of light is closer to the eye than the point of reversal, this can only be effected by the concave mirror, which, therefore, has so much advantage over the plane mirror. It is of some practical importance in a few cases, in which the aberration invades the visual zone.

With *negative aberration* the advantage lies with the plane mirror, but is of still less practical importance on account of the smaller number of cases of aberration of this kind.

With a concave mirror, the distance from which it can be used with advantage is fixed, and the surgeon being able to readily check his position by a mark on the neighboring wall, or some similar device, there is no need of the measurement of the distance between the surgeon and patient; but all changes of the movement of the light in the pupil must be effected by a change of the lens before the eye. On the other hand, the plane mirror can be used from any fixed distance, but allows also a variation of the distance of the surgeon from the patient, and, therefore, requires the fewer changes of the lens before the eye.

This latter advantage of the plane mirror over the concave mirror may not seem great; but when repeated inspections of the light movement are required it is important. The disadvantage of the concave mirror may be lessened by using a lens-disk; Yet with it the complete break between the appearances brought out by successive lenses makes the imformation they give less satisfactory than that obtained by movement of the surgeon's eye, with which the successive appearances of light and shade pass gradually into one another. Perhaps the greatest advantage of the plane over the concave mirror is that it thus facilitates the study of aberration and irregular astigmatism.

INDEX.

ABERRATION, 16, 36, 41, 43, 55, 59, 63, 82, 96, 105
Accommodation,...........84, 97, 104
Accuracy, ...35, 39
Advantages,........................5, 105
Apparatus,................12, 35, 69, 98
Apparent movement, 21, 22, 25, 28, 29, 49, 53
Appearances in pupil, 13, 30, 46, 56
Application,......................69, 86
Area of light,...21, 23, 24, 30, 47, 53
Artificial eye......................17, 68
Astigmatism, 7, 16, 36, 40, 45. 56, 68, 74, 91, 96, 105
Atropin 104
Axonometer 103

BAND appearance, 7, 12, 31, 40, 46, 48, 53, 76, 92, 95
Bowman,........................7, 10, 46
Brightness of light,..............32, 36
Burnett,........................102, 104

CHARNLEY,......................9
Chibret 9. 10, 11
Concave mirror, 9, 23, 26, 39, 41, 43, 50, 53, 60, 62, 87, 105
Conditions of accuracy,...........35
Conical cornea,....................7, 64
Contents,..........................3
Couper,........................8, 102
Cuignet,........................8, 9, 10

DARK ROOM,....................35
Daturin,.........................104
Derby,............................35
Difficulties,.................11, 38, 70
Dioptric scale....................103
Dioptroscopie...................11
Direction,........46, 47, 53, 76, 91, 95
Distance, 38, 42, 43, 44, 48, 51, 87, 103, 106
Donders,.........................7, 8
Doyne, ...102
Duboisin,.........................104

EGGER, ...11
Emmetropia,.......23, 26, 74, 91
Enlargement,..................29, 48

Erect image,............18, 25, 59, 88
Extra-visual zone...................58

FACIAL light area.............23, 25
Fantoscopie......................10
Focal lengths,....................103
Forbes,....................8
Form of light area, 30, 46, 49, 55, 59, 65, 66
Fundus-reflex test...................11

GALEZOWSKI........8, 11
General principles,...........18

HAINES,........................102
Hartridge, ...11
History,.........................7
Homatropin,....................104
How to study.......12, 49, 66, 68, 84
Hyoscyamin, ...104
Hyperopia,.............7, 23, 26, 70, 88

ILLUMINATION, 21, 32, 35, 56, 60
Illustrations, 19, 22, 24, 29, 31, 47, 48, 54, 56, 60, 61, 65, 67, 69, 82, 100, 102, 103
Immediate source,............21, 37, 39
Inclinometer,--......................103
Indicators........................103
Inverted image,...........18, 25, 88
Irregular astigmatism, 7, 40, 42, 43, 55, 82, 96

JACKSON, 9, 10, 32, 79, 102, 104
Juler,..............................9

KERATOSCOPIE,.................10
Koroscopie,....................11

LAMBERT,........................102
Landolt..........................11
Learning the test 13
Lenses,.......................101, 102
Light area, 21, 23, 25, 26, 30, 32, 41, 49, 55, 59, 65, 67
Light source,...21, 22, 35, 37, 69, 87

MAGNIFICATION of retina, 29, 48
Mengin,.........................8

(107)

INDEX.

Meridians, ...45, 76, 79, 95, 103
Mirror,........ ....35, 70, 98
Morton,............... 9
Movements of light, 18, 21, 23, 24, 27, 28, 51, 53. 62, 64, 67, 88
Mydriatics................................104
Myopia,.............7, 18, 23. 26 72, 89

NAME,....................................10
  Near point,.......................84
Negative aberration, 42, 57, 62, 105

OBLIQUITY of lens,..... 68
  Oliver, ...11
Optical principles.....................18
Original source of light,...21, 23, 38

PARENT,...................8, 9, 10, 11
  Perkins............................102
Plane mirror, 9. 22, 26. 38, 39, 41, 43, 50, 51, 59, 62, 69, 105
Point of reversal,....20, 33, 39, 45, 47
Position,.........12, 39, 42, 48, 69, 87
Positive aberration,........41, 59, 105
Practical application,.......13, 69, 87
Preface, ...5
Prince,..................................103
Principal meridians, 45, 76, 79, 95, 103
Pupil, size of,.........................105
Pupillary shadows, 25, 30, 32, 46, 55, 59, 66
Pupilloscopie................ 11

RANDALL,........ 42
  Rate of movement,...27, 30. 64
Real movement,...21, 22, 23, 24, 28

Regular astigmatism, 7, 16, 36, 40, 45, 74, 91, 96, 105
Retinal area, 22, 23, 25, 27, 30, 37, 41, 51
Retinal enlargement,............29, 48
Retinophotoscopie,.....................11
Retinoscopy,..........................10
Retinoskiascopie.......................11
Reversal,......... ....18, 33, 45, 47

SCISSORS movement,............66
  Scopolamin,...... 104
Shade,...................35, 69, 87, 100
Shadows,........24, 30, 46, 55, 59, 66
Shadow-test,...........................10
Sight-hole,...................35, 37, 98
Size of mirror,......... 99
Skiascopy,........... 11
Smith, Priestly,.........................10
Source of light, 21, 22, 35, 37, 69, 87
Story,............. 9
Study of test,.........12, 49, 66, 68, 84
Sureau..................................102
Symmetrical aberration, 41, 43, 57, 59, 61, 62

THORINGTON,............100, 103
  Tait............................102

UMBRASCOPY,. 11
  Use of test..................13, 15

VISUAL ZONE,..........43, 58, 82

WEILAND,........... ....32, 79
  Würdemann..................102

www.ingramcontent.com/pod-product-compliance
Lightning Source LLC
Chambersburg PA
CBHW020145170426
43199CB00010B/900